Solid State Quantum Information

An Advanced Textbook

Quantum Aspect of Many-Body Systems

Wonmin Son • Vlatko Vedral
NUS, Singapore *University of Oxford, UK*

Solid State Quantum Information

An Advanced Textbook

Quantum Aspect of Many-Body Systems

World Scientific
NEW JERSEY • LONDON • SINGAPORE • BEIJING • SHANGHAI • HONG KONG • TAIPEI • CHENNAI • TOKYO

Published by

World Scientific Publishing Europe Ltd.

57 Shelton Street, Covent Garden, London WC2H 9HE

Head office: 5 Toh Tuck Link, Singapore 596224

USA office: 27 Warren Street, Suite 401-402, Hackensack, NJ 07601

Library of Congress Cataloging-in-Publication Data
Names: Vedral, Vlatko, author. | Son, Wonmin, 1971– author.
Title: Solid state quantum information : an advanced textbook : quantum aspect of many-body systems / by Vlatko Vedral (Oxford), Wonmin Son (NUS, Singapore).
Description: New Jersey : World Scientific, 2018. | Includes bibliographical references and index.
Identifiers: LCCN 2017053012 | ISBN 9781848167643 (hc : alk. paper)
Subjects: LCSH: Solid state physics--Textbooks. | Many-body problem--Textbooks. | Quantum theory--Textbooks.
Classification: LCC QC176 .V43 2018 | DDC 530.4/1--dc23
LC record available at https://lccn.loc.gov/2017053012

British Library Cataloguing-in-Publication Data
A catalogue record for this book is available from the British Library.

For any available supplementary material, please visit
http://www.worldscientific.com/worldscibooks/10.1142/P797#t=suppl

Desk Editors: Suraj Kumar/Jennifer Brough/Koe Shi Ying

Typeset by Stallion Press
Email: enquiries@stallionpress.com

Printed in Singapore

For our friends and our family

Preface

This book came from a lecture course on "Advanced Solid State Physics" at the Physics Department of NUS Physics between 2008 and 2012. The course was about 30 lectures long. It is given together with a number of additional problem solving sessions that were meant to aid the lectures and expand on some of the key concepts. The course had positive response from the students (about 40 attended it each year). We thought then that it would be great to turn the lecture notes into a textbook. The profitable niche we spotted comes from the fact that the solid-state physics plays an important role in quantum information and computation, but that there are virtually no textbooks on that subject at the advanced undergraduate level.

At the same time, the study of entanglement offers a new and fresh perspective on some more traditional solid-state topics, such as that of phase transitions and criticality. Not only is the solid-state seen as the most promising technology for quantum computers, but also quantum computers themselves offer an exciting possibility to simulate more complex solid-state systems and possibly help us solve some open problems, such as the origin of high temperature superconductivity. This symbiotic relationship between the two topics forms the cornerstone of this book.

The book therefore presents some traditional solid-state topics in the first two chapters but from a new perspective with basic understanding of quantum & statistical mechanics that the reader will hopefully find entertaining and original. Chapters 3 and 4 then venture into the exciting areas of quantum information in the solid state. Every chapter contains a number of problems that the reader might find useful in testing their understanding of the topic. Some of the problems are deliberately designed to be more challenging and give the reader a taste of what research in this area would be like.

We have very much enjoyed collaborating on this book and hope that the reader will like our style and approve of our approach to this exciting subject. In addition, we would like to greatly acknowledge Donggun Lee, who is a research PhD student in Son's group at Sogang University. This book could not have been completed without his supports. We also would like to thank all the people who have encourage our effort as it could be concluded with this fruitful outcome.

W. Son & V. Vedral

About the Authors

Wonmin Son is a faculty at Sogang University, Seoul, Korea, and visiting professor at the University of Oxford. He has published more than 30 research papers garnering roughly over 2000 citations and works on the various topics in the field of quantum optics, quantum information and foundational physics. Previously, he had been selected as a British Council Chevening Scholar in 2001 and has been appointed as an Erwin Schrodinger junior research fellow in 2007.

Vlatko Vedral is a professor of physics at the University of Oxford and also at the National University of Singapore (where he is a PI at the Centre for Quantum Technologies). He has published over 300 research papers (a significant fraction of which is in premier journals such as *Nature*, *Physical Review Letters* and *Reviews of Modern Physics*) on various topics in quantum physics and quantum computing. He has given numerous invited plenary and public talks in the last 25 years of his career. These include a specialized talk at a Solvay meeting (2010) and a popular one at the International Safe Scientifique (2007). He was awarded the Royal Society Wolfson Research Merit Award in 2007 and the World Scientific Medal and Prize in 2009.

Contents

Chapter 1

Introduction

Solid-state physics, the largest branch of condensed matter physics, is the study of rigid matter through scientific methods such as crystallography, metallurgy, electromagnetism and quantum mechanics. Solid-state physics considers how the large-scale properties of solid materials result from their atomic-scale properties. Thus, the solid-state physics is the theoretical basis of material science, and it has direct application in the various fields, for example, the technologies for transistors and semiconductors. To understand all those, it is crucial to use quantum mechanics and, in particular, quantum statistics. At first, we discuss basic methods to investigate the structure of a solid in this chapter. At the same time, the basic axioms of quantum theory as well as statistical mechanics will be briefly mentioned here.

1.1. Basic Structures of Solid

The solid materials are formed by densely packed atoms whose electrical, magnetic and thermal properties are determined by the way that they are arranged. We present the way to study the physical properties of solids. Atomic structure of solid is important factor which determines its large-scale properties. For example, spatial dimensionality of the solid determines the structure of its physical behavior.

A possible method to investigate the structure of a crystal is made through the X-ray beam which is scattered from solids. The basic idea was firstly proposed by German physicist von Laue in 1912. By looking at the intensity pattern of scattered X-ray, the structure of the solid can be

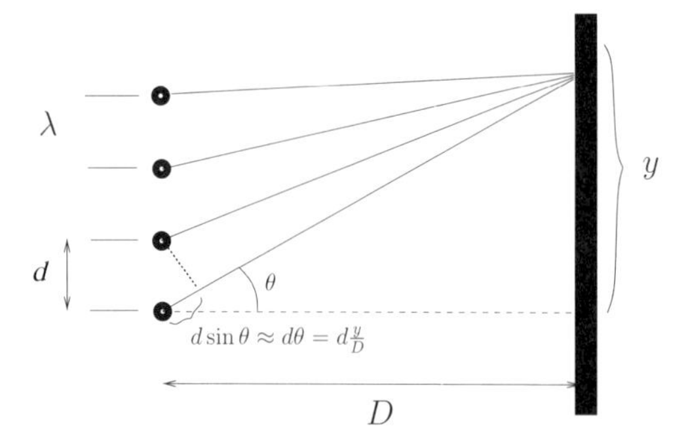

Fig. 1.1. Lattice Interference: The plane wave incident to the atoms in an array creates the diffraction patterns.

revived. The logic is that if plane wave is incident to atoms in an array, then the atoms act as a multiple slit and create diffraction grating patterns as shown in Fig. 1.1.

Due to the difference of the path length, the electromagnetic wave interferes each other. The wave function ψ of the field at the vertical position x varies

$$\psi = A_0[e^{ikx} + e^{ik(x+d\theta)} + e^{ik(x+2d\theta)} + \cdots + e^{ik(x+Nd\theta)}]. \tag{1.1}$$

Therefore, the intensity of the field becomes

$$I = |\psi|^2 \sim |A_0|^2 \cdot \left|\frac{1 - e^{ikNd\theta}}{1 - e^{ikd\theta}}\right|^2 \sim \frac{\sin^2(\frac{kNdy}{2D})}{\sin^2(\frac{kdy}{2D})}. \tag{1.2}$$

The profiled intensity tells us about the probability amplitude coincide with the periodicity of the grating. Following the von Laue's intuition [1], the same should work for any solids.

Typical spacing between atoms in a solid is 1–10 Å. Therefore, to inspect the structure of a solid using scattering techniques, one needs a probe beam with wavelength $\lambda \leq 10^{-9}$ m. The order of wavelength of X-rays is 10–10^{-2} nm, that of neutrons is 1–10 Å and that of electrons is 10^{-2} pm.

In a real situation, the lattice of a solid is layered in the two-dimensional (2D) space so that the interference pattern of scattered field is more complicated than the idealized one above. Bragg diffraction occurs when the wavelength λ of electromagnetic radiation incident upon a crystalline sample is comparable to atomic spacings. For a crystalline solid, the radiation is scattered in a specular fashion by the atoms in the system and the wave undergo constructive interference in accordance to Bragg's law. When the waves are scattered from lattice planes separated by the distance d, the scattered waves interfere constructively when they remain in phase if the path length of each wave is equal to an integer multiple of the wavelength. The condition for the path difference between two waves undergoing constructive interference is given by

$$2d \cdot \sin\theta = n\lambda, \tag{1.3}$$

where θ is the scattering angle and n is integer. The formula state Bragg's law which describes the condition for constructive interference from successive crystallographic planes of the crystalline lattice.

The 3D lattice plane can be described by three integer numbers called Miller indices (h, k, l).[a] Miller indices are a notation system in crystallography for the solid planes and its directions in crystal lattices. Each index denotes a plane orthogonal to a direction (h, k, l) in the basis of the reciprocal lattice vector $\vec{G}$. Due to the momentum conservation, the reciprocal lattice vector is obtained through the difference between the incoming and outgoing wave vectors,

$$\vec{G} = \vec{k}_{\text{out}} - \vec{k}_{\text{in}}, \tag{1.4}$$

where $\vec{G} = h\vec{a} + k\vec{b} + l\vec{c}$ and $\vec{a}$, $\vec{b}$, $\vec{c}$ are the vector components of the lattice. It also can be used to derive Bragg law (1.3) as it represents the lattice space. The space distance d between the lattice planes can be related with the Miller indices

$$d = \frac{a}{\sqrt{h^2 + k^2 + l^2}}, \tag{1.5}$$

where a is the atomic lattice spacing of cubic crystal. Combining the relation with the Bragg scattering condition,

$$\left(\frac{\lambda}{2a}\right)^2 = \frac{\sin^2\theta}{h^2 + k^2 + l^2} \tag{1.6}$$

would give the selection rule of scattering beam for different crystals. Thus for all sets of Miller indices there is an angle that will satisfy the Bragg condition such that the value of $\frac{\lambda}{2a}$ is a constant.

The important usage of the structural constants is to give the distinction between the crystal structures. The structures are identified

[a] The definition of Miller index is the x, y, z-directional scales by integer number.

by the different ways of stacking basic constituent, atoms in this case. In the ideal case, possible structures are the simple cubic (SC), the body-centered cubic (BCC) and the face-centered cubic (FCC) crystal structures as they are illustrated in Fig. 1.2. The discrimination can be made through the comparison between the allowed Miller indices and the properties can actually be evaluated from the analysis of the X-ray spectrum.

Let us discuss the physical conditions when the X-ray scattering occurs. If X-ray is triggered to the target crystal, the beam is reflected at a particular angle. At the incidence, the vector perpendicular to the reflective surface is defined by $\vec{G}$. The relationship between $\vec{k}$ for the beam and $\vec{G}$ for the surface satisfies the momentum conservation law unless the scattering process is absorptive. From the geometric construction in Fig. 1.3, the relation $\Delta\vec{k} = \vec{k}_{\text{out}} - \vec{k}_{\text{in}} = \vec{G}$ provides the way to analyze the interference pattern of the reflected beam. The diffraction condition implies that if the reflective wave vector $\vec{k}_{\text{out}}$ and the incident wave vector $\vec{k}_{\text{in}}$ are known, then

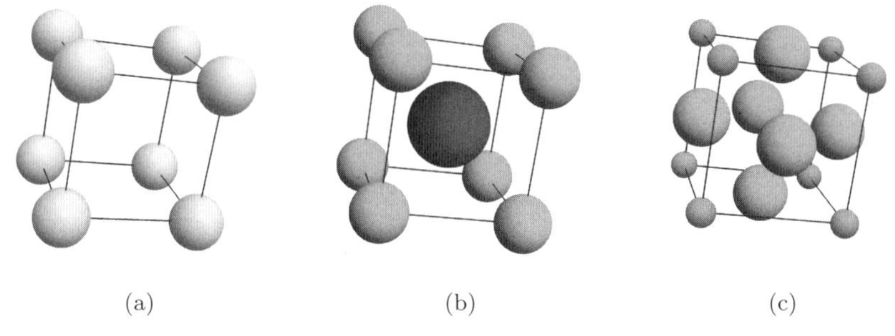

Fig. 1.2. Three cubic lattice structures: (a) the simple cubic (SC), (b) the body-centered cubic (BBC) and (c) the face-centered cubic (FCC).

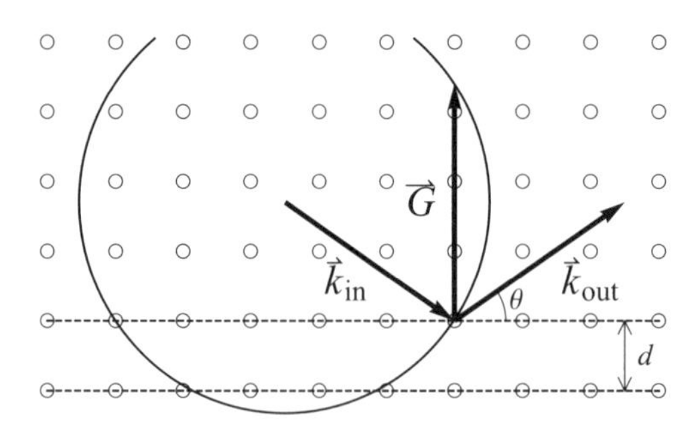

Fig. 1.3. Bragg scattering: The reflected and diffracted beams lie on the surface of propagating sphere which is called Ewald sphere.

it is possible to find the reciprocal lattice vector $\vec{G}$ as we can reconstruct the lattice structure.

Since the solid is made out of the large collection of atoms, the structural property of the solid can be discussed only after the detailed inspection of the quantum mechanical characteristics of the individual atoms. For the purpose, we will begin with the basic description of quantum mechanics and the statistical treatment of the large systems in the following sections.

1.2. Basic Formulation of Quantum Mechanics for a Simple System

The description of quantum mechanics is made through the series of rules to evaluate the probabilities. The probability in a quantum system is for the events when an experimental test on a microscopic scaled object has been performed.

In the theory, the fundamental principle that the quantum objects satisfy, e.g. particle-wave duality and Heisenberg uncertainty relation, is expressed through the distribution of a particle in the space as it is called the wave function. They follow basic postulates in quantum mechanics and the procedure of a quantum test can be formulated into three different operational stage as to be illustrated as below.

- Compact formulation of quantum mechanics:
  - *States*: The preparation procedure of a given quantum state is described in terms of a complex vector having unit norm and the quantum state is called pure state. When the state is in a 2D vector space representing only two possible mode of existence, it can be written, in Dirac notation, as

$$|\psi\rangle = \alpha|0\rangle + \beta|1\rangle \equiv (\alpha, \beta), \tag{1.7}$$

where $|0\rangle$ and $|1\rangle$ are the basis of the orthonormal vectors. Therefore, α and β are arbitrary complex numbers satisfying unit norm $|\alpha|^2 + |\beta|^2 = 1$. The vector notation is possible to be extended further when the system is subjected under more than two degree of freedom.

When the vector is projected into a space of particular spatial distribution, it becomes a function that represents the distribution of wave-like spreading for the objects as it is given $\psi(r) = \langle r|\psi\rangle$. The wave function $\psi(r)$ provides the quantum state distribution over the allowed position r and the position can be 3D $\vec{r}$ in general.

If the system is in the statistical mixture of several different pure states, the state should be written by positive operator, in mathematical term, called density operator or Hermitian operator. Such a mixture can be obtained when one starts from the arbitrary composite two-level states A and B under the total superposition as

$$|\psi\rangle_{AB} = \alpha_1|00\rangle_{AB} + \alpha_2|01\rangle_{AB} + \alpha_3|10\rangle_{AB} + \alpha_4|11\rangle_{AB}. \tag{1.8}$$

When the access to the second system is impossible, the one of the part should be traced out and the state remained to be in a statistical mixture as

$$\begin{aligned}\mathrm{Tr}_B[|\psi\rangle_{AB}\langle\psi|] &= (|\alpha_1|^2 + |\alpha_2|^2)|0\rangle\langle 0| + (|\alpha_3|^2 + |\alpha_4|^2)|1\rangle\langle 1| \\ &\quad +(\alpha_1\alpha_3^* + \alpha_2\alpha_4^*)|0\rangle\langle 1| + (\alpha_3\alpha_1^* + \alpha_4\alpha_2^*)|1\rangle\langle 0| \\ &= \hat{\rho}_A,\end{aligned} \tag{1.9}$$

where the density matrix $\hat{\rho}_A$ satisfies the Hermitian condition $\hat{\rho}_A = \hat{\rho}_A^\dagger$. In general, a quantum state is represented by a Hermitian operator and becomes a pure state at the limit of rank one projection as $\alpha_2 = \alpha_3 = \alpha_4 = 0$ for example.

- *Evolution*: Time evolution of a quantum state is subject to the Schrödinger equation under the potential $V(r)$,

$$\left[-\frac{\hbar^2}{2m}\nabla^2 + V(r)\right]\psi(r) = E\psi(r). \tag{1.10}$$

When the system is evolved without energy loss as a constant of energy E, the dynamics is given by unitary operator $U = \exp[itH]$. Mathematically, the operator U satisfies the unitarity condition $UU^\dagger = U^\dagger U = \mathbb{1}$ while $H = -\frac{\hbar^2}{2m}\nabla^2 + V(r)$ is Hermitian operator $H^\dagger = H$.

- *Measurements and observables*: Probability of an event in a quantum test is dependent upon the type of measurement performed to the state. Following the Born rule, measuring action is represented by the complete set of state vectors and the probability of obtaining certain outcome, say $|\phi_i\rangle$, in the measurement of a state $|\psi\rangle$ is given by the norm of inner product of the state vectors as

$$|\langle\phi_i|\psi\rangle|^2 = \text{Probability of the } i\text{th outcome} = p_i. \tag{1.11}$$

Quantum systems are normally very sensitive to any disturbance so that measuring action also changes the state of a system and makes the state projected into a state of measurement direction. Two incompatible measurement on a single system cannot be performed simultaneously with arbitrary precision. Observable describes the physical variable to be measured and it is averaged value of the ensemble measurement as it is in the form,

$$\langle\hat{A}\rangle = \sum_i p_i A_i = \langle\psi| \left(\sum_i A_i |\phi_i\rangle\langle\phi_i| \right) |\psi\rangle. \tag{1.12}$$

When the physical variable is real-valued, the observable is given by Hermitian operator $\hat{A} = \sum_i A_i|\phi_i\rangle\langle\phi_i|$ with real-valued number A_i for all i.

- *Fermion and bosons*: In the many-body physics, statistics of an identical particle results in physically distinctive particle descriptions. The particles are in the mode to satisfy the rule of symmetrization as the total state remains either unchanged or differ by a certain phase factors under the particle exchanges. Depending upon the phase difference, the state either called as boson or fermion. Photon and phonon are bosonic particles and electron is the best known fermionic particles. We would discuss the symmetries in the wave function in detail in Chapter 2.

1.3. Statistical Mechanics and Partition Function

When full knowledge about individual particles in an ensemble is not available, it is not possible to know the exact equation of motion for the total system. In that scenario, the royal route to calculate quantum mechanics can be provided by statistical mechanics. By knowing the Hamiltonian of the system, one can derive all the macroscopic properties of the statistical ensemble using partition function.

In statistical mechanics, the partition function, Z, encodes the statistical properties of a system in thermodynamic equilibrium. It is a function of temperature and other parameters, such as the volume enclosing a gas. Most of the aggregate thermodynamic variables of the system, such as the total energy, free energy, entropy, and pressure, can be expressed in terms of the partition function or its derivatives.

There are actually several different types of partition functions, each corresponding to different types of statistical ensemble (or, equivalently, different types of free energy). When the system is in a completely isolated circumstance, the partition function is to be described by microcanonical ensemble. In that case, all the thermodynamic variables are invariant and constant over the time. Contrarily, the canonical partition function applies to a canonical ensemble, in which the system is allowed to exchange heat with the environment at fixed temperature, volume, and number of particles. The grand canonical partition function applies to a grand canonical ensemble, in which the system can exchange both heat and particles with the environment, at fixed temperature, volume and chemical potential. Other types of partition functions can be defined for different circumstances; see partition function (mathematics) for generalizations.

1.3.1. *Microcanonical ensemble*

In microcanonical ensemble, it is assumed that the values of the particle number N, the volume V and the energy E are the constant over the time. As the second law of thermodynamics applies to isolated systems, we investigate it as the first case. The microcanonical ensemble describes an isolated system as the entropy of such a system can only increase. It leads to the system that the maximum of its entropy corresponds to an thermodynamic equilibrium state for the system.

Because an isolated system keeps a constant energy, the total energy of the system does not fluctuate. Thus, the system can access only those of its microstates that correspond to a given value "E" of the energy. The internal energy of the system is then strictly equal to its energy.

Let us call $\Omega(E)$ the number of microstates corresponding to this value of the system's energy. The macroscopic state of maximal entropy for the system is the one in which all microstates are equally likely to occur, with probability $1/\Omega(E)$, during the system's fluctuations.

$$S = -k_B \sum_{i=1}^{\Omega(E)} \left\{ \frac{1}{\Omega(E)} \ln \frac{1}{\Omega(E)} \right\} = k_B \ln(\Omega(E)), \tag{1.13}$$

where S is the system entropy, and k_B is Boltzmann's constant.

The partition function can be used to find the expected (average) value of any microscopic property of the system, which can then be related to macroscopic variables. For instance, the expected value of the microscopic energy E is "interpreted" as the microscopic definition of the thermodynamic variable, for instance internal energy U, and the value can be obtained by taking the derivative of the partition function with respect to the temperature. Indeed,

$$\langle E \rangle = \frac{\sum_i E_i e^{-\beta E_i}}{Z} = -\frac{1}{Z}\frac{dZ}{d\beta}$$

implies, together with the interpretation of $\langle E \rangle$ as U, the following microscopic definition of internal energy:

$$U: = -\frac{d \ln Z}{d\beta}.$$

The entropy can be calculated by (see Shannon entropy)

$$\frac{S}{k} = -\sum_i p_i \ln p_i = \sum_i \frac{e^{-\beta E_i}}{Z}(\beta E_i + \ln Z) = \ln Z + \beta U,$$

which implies that

$$-\frac{\ln(Z)}{\beta} = U - TS = F$$

is the thermodynamic free energy of the system or in other words,

$$Z = e^{-\beta F}.$$

Having microscopic expressions for the basic thermodynamic potentials U (internal energy), S (entropy) and F (free energy), derived from the partition function, is sufficient to obtain the expressions for other thermodynamic quantities. The basic strategy is as follows. There may be an intensive or extensive quantity that enters explicitly in the expression for the microscopic energy E_i, for instance magnetic field (intensive) or volume (extensive). Then, the conjugate thermodynamic variables are derivatives of the internal energy. The macroscopic magnetization (extensive) is the derivative of U with respect to the magnetic field, and the pressure (intensive) is the derivative of U with respect to volume (extensive).

The treatment of the macroscopic system in this section assumes no exchange of matter (i.e. fixed mass and fixed particle numbers). However, the volume of the system is variable which means the density is also variable. This probability can be used to find the average value, which corresponds to the macroscopic value, of any property, J, that depends on the energetic state of the system by using the formula:

$$\langle J \rangle = \sum_i p_i J_i = \sum_i J_i \frac{e^{-\beta E_i}}{Z},$$

where $\langle J \rangle$ is the average value of property J. This equation can be applied to the internal energy, U:

$$U = \sum_i E_i \frac{e^{-\beta E_i}}{Z}.$$

Subsequently, these equations can be combined with known thermodynamic relationships between U and V to arrive at an expression for pressure in terms of only temperature, volume and the partition function. Similar relationships in terms of the partition function can be derived for other thermodynamic properties.

1.3.2. *Canonical ensemble*

When the system is in contact with a head bath at constant temperature, the energy of the systems in the assembly is not identical from the microcanonical ensemble. The basic assumption is that a thermodynamically large system is in constant thermal contact with the environment, with a temperature T, and both the volume of the system and the number of constituent particles fixed. The kind of system is called a canonical ensemble. Let us label with i ($i = 1, 2, 3, \ldots$) the exact states (microstates) that the system can occupy, and denote the total energy of the system when it is in microstate i as E_i. Generally, these microstates can be regarded as analogous to discrete quantum states of the system. In the circumstance, the statistical distribution of a system in a thermal equilibrium follows Boltzmann statistics and the partition function of the given system

$$Z = \sum_i e^{-\beta E_i} \quad \text{where } \beta = \frac{1}{kT} \tag{1.14}$$

allows us to evaluate free energy

$$F = -kT \ln Z. \tag{1.15}$$

In order to demonstrate the usefulness of the partition function, let us calculate the thermodynamic value of the total energy. This is simply the expected value, or ensemble average for the energy, which is the sum of the microstate energies weighted by their probabilities:

$$\langle E \rangle = \sum_s E_s P_s = \frac{1}{Z} \sum_s E_s e^{-\beta E_s} = -\frac{1}{Z} \frac{\partial}{\partial \beta} Z(\beta, E_1, E_2, \ldots) = -\frac{\partial \ln Z}{\partial \beta}$$

or, equivalently,

$$\langle E \rangle = k_B T^2 \frac{\partial \ln Z}{\partial T}.$$

Incidentally, one should note that if the microstate energies depend on a parameter λ in the manner

$$E_s = E_s^{(0)} + \lambda A_s \quad \text{for all } s,$$

then the expected value of "A" is

$$\langle A \rangle = \sum_s A_s P_s = -\frac{1}{\beta} \frac{\partial}{\partial \lambda} \ln Z(\beta, \lambda).$$

This provides us with a method for calculating the expected values of many microscopic quantities. We add the quantity artificially to the microstate energies (or, in the language of quantum mechanics, to the Hamiltonian), calculate the new partition function and expected value, and then set "λ" to zero in the final expression. This is analogous to the source field method used in the path integral formulation of quantum field theory.

Using the free energy, all relevant macro-observables also can be followed. For example, internal energies of the magnetization M and magnetic susceptibility χ of a system are

$$M = \frac{\partial F}{\partial B}, \quad \chi = \frac{\partial^2 Z}{\partial B^2}, \quad U = \frac{\partial F}{\partial T}, \tag{1.16}$$

which are all measurable macroscopic quantities. The canonical ensemble becomes microcanonical ensemble when the relative fluctuation of the energy value becomes quite negligible. For all practical purposes, a system in the canonical ensemble has an energy equal to the mean energy same as in the microcanonical ensemble.

1.3.3. *Grand canonical ensemble*

Grand canonical ensemble is the statistical distribution of the system when the canonical ensemble is allowed to exchange the particle between the system and the bath. Due to the possible fluctuation of the particle number and the energy, the partition function is need to be defined with the statistical distribution of the particle number. When the system and the reservoir have a common temperature T and a common chemical potential μ, the partition function of the grand canonical ensemble becomes

$$Z_G = \sum_{N=0}^{\infty} e^{-\alpha N} Z, \tag{1.17}$$

where $\alpha = -\mu/kT$ and $Z = \sum_i e^{-\beta E_i}$ is the partition function for a fixed particle number.

From the partition function, the expectation value of the particle number and the energy can be obtained

$$\langle N \rangle = -\frac{\partial}{\partial \alpha} \ln Z_G \quad \text{and} \quad \langle E \rangle = -\frac{\partial}{\partial \beta} \ln Z_G. \tag{1.18}$$

In the derivation of the thermodynamic quantities, all the relations are not fundamentally different and the treatment of the partition function is similar. In the ensemble, the Helmholtz-free energy and the entropy are given by

$$F = N\mu - PV = -kT \ln\left(Z_G/e^{-\alpha N}\right) \quad \text{and} \quad S = (E-F)/T \tag{1.19}$$

and the relevant quantities can be derived from them as discussed above.

1.4. Problems

1. **CBCC versus FCC:** Compare the packing density of atoms in face-centered cubic (FCC) and body-centered cubic (BCC) crystal structures. Which structure would you expect to be favored by Nature and why?
2. **Structure factor:** Structure factor $S(\vec{G})$ describes all possible scattered lights and $|S(\vec{G})|^2$ is proportional to the scattering intensity. From its definition, $S(\vec{G}) = \sum_j f_j e^{2\pi i (h x_j k y_j + l x_j)}$ where j means jth atom in unit cell. From the relation, show that the scattering intensity exists only for even $h+k+l$ in the BCC crystal (e.g. a pure Cs crystal).
3. **Schrödinger equation in the free space:** Let us consider a quantum particle in a free space without constraint as $V(r) = 0$.

   (a) Solve the Schrödinger equation (1.10) for the particle in 1D box distanced L and obtain the normalized form of the wave function.

(b) There are two possible solutions of the equation above and they can be expressed as $\psi_k(x)$ and $\psi_{-k}(x)$. Show that they are orthogonal to each other.

(c) Show that the linear combination of the two solutions is also the solution of Eq. (1.10):

$$\psi(x) = \alpha\psi_k(x) + \beta\psi_{-k}(x) \tag{1.20}$$

and discuss the relation between the vector representation of a quantum state and the wave function of a quantum particle.

Chapter 2

Electrical Conductivity in Solid

Electrical conductivity is a measure of a material's ability to conduct an electric current. When an electrical potential is placed across a conductor, its movable charges flow is giving rise to an electric current. Solids that can carry the electric current have atoms with outer electrons and the electrons are free to move throughout the entire length of the material. The phenomena of electrical conductivity have explained by several different models and they had provided various levels of understanding. They had applied classical, semiclassical theories and theories with first quantized and second quantized versions.

We start with simplest classical model of electrical conductivity which is still good for basic understanding. However, the theory is limited so that they had failed to explain the other phenomena in a solid and it had been overcome by more elaborated theories. It is quantum theory which has successfully explained them with consistency.

2.1. Classical (Drude) Model

In 1900, Paul Drude proposed a model of electrical conductivity in a metallic material. The model explains the transport properties of electrons in metals with an assumption that the microscopic behavior of electrons in a solid may be treated as classical motion of point particles. The dynamics of electron resembles a pinball machine as like sea of balls bouncing and rebouncing in the lattices that is composed of heavier, relatively immobile positive ions. In the model, several assumption had been made for the matter of simplification and they are as follows:

(i) Electrons are classical particles.
(ii) They only collide with ions as they move.
(iii) Between collision they do not interact with each other or ions.
(iv) Collision is made instantaneously with some probability per unit time.
(v) Thermal equilibrium only achieved in this way.

In that circumstance, the current density j of electrons with charge e and average velocity v is

$$j = -nev = -\frac{ne}{m_e}p, \tag{2.1}$$

where m_e is the electron mass and n is the electron density. Here p is the average momentum of the electron. During the time interval δt, the momentum changes of electrons are caused due to the electron–atom collision. The probability of a collision is $\delta t/\tau$ so that the momentum change of electron which does NOT suffer from the collision is

$$p(t+\delta t) = \left(1 - \frac{\delta t}{\tau}\right)\left[p(t) + f(t)\delta t\right], \tag{2.2}$$

where τ is average collision time and $f(t)$ is the force acting on the electrons during the collision. Therefore, at an infinitesimally short time limit, the equation for the momentum changes becomes

$$\frac{dp(t)}{dt} = -\frac{p(t)}{\tau} + f(t), \tag{2.3}$$

where the contribution from electrons that have collided with being on the order of $(dt)^2$ is ignored. It tells us that the collisions produce a frictional damping term in the momentum changes.

The electrical conductivity σ is defined as the ratio between the applied electric field E and the electric current J such that

$$J = \sigma E \qquad \text{(by definition)}, \tag{2.4}$$

where the electric field is given by the amount of force exerted in an electron at a position $F = -eE$. Therefore, at a steady state where $\frac{dp}{dt} = 0$, it is possible to derive the electric conductivity from (3.7), (2.3) and (2.4) and it reads

$$\sigma = \frac{ne^2\tau}{m_e}. \tag{2.5}$$

If we substitute the room temperature value for typical metals, $\frac{1}{\sigma} \approx 10\,\mu\Omega\text{cm}$ [2] along with a typical $n \approx 10^{22}$–$10^{23}\,\text{cm}^{-3}$, a value for the average collision time $\tau \approx 10\,\text{fs}$ is obtained. The value of the average collision time is well agreed with data for the atomic structure of typical metal. In this picture, the electrons can be considered as the particles of classical gas. For an ideal gas, the average velocity can be derived from the mean kinetic energy as $U_K = m_e\langle v^2\rangle/2 = 3k_BT/2$. From the average velocity of the electron with the average collision time, the mean free path is given as $l \sim 0.1$–$1\,\text{nm}$ which is roughly same as the interatomic distance in metals. This is consistent with what Drude picture of electrons in a metallic material predicts.

However, the model is failed when it is used to explain the other physical properties, for example the heat conductivity in a metal. If the electrons are treated as a classical ideal gas, then their heat capacity is

$$C_{el} = \frac{du}{dT} = \frac{3}{2}nk_B, \tag{2.6}$$

where $u = nU_K$ is the total energy of gas with density n. Based on the Eq. (2.6), the heat capacity of a metal is independent of temperature. Experimentally, it is known that the low temperature heat capacity follows the relationship as

$$c_v = \gamma T + \alpha T^3, \tag{2.7}$$

where T is the temperature [3]. Based on the Debye's model, the contribution of T^3 term is from phononic part which is the vibrational mode of the ions in the solid. It leads us to suspect that the term with the scale T is from the electronic part. Taking only the electronic part, the value of heat capacity is two orders at magnitude higher than what the Drude model assumed at room temperature T [4]. Thus, the conclusion drawn from the Drude model that the heat capacity is independent of temperature is not correct. Although the model was flawed, its prediction of the correction order of magnitude for the Wiedemann–Franz ratio [5] provoked a large amount of further work which lead to the Sommerfeld model.

2.2. Semiclassical (Sommerfeld) Model

If one assume that electrons are free to move anywhere in the solid, within the boundary, then their energies are quantized according to quantum mechanics and Pauli exclusion principle. The quantization rule of electron tells us that the number of electrons is discrete. In addition, according to Pauli's exclusion principle, different momenta k of the propagating electrons are canceled so that only one electron can occupy the single momentum space.

In general, the potential energy of the electrons remains uniform throughout the solid. The force of repulsion between electrons and the force of attraction between electrons and lattice ions can be neglected. For electron densities, a Fermi–Dirac distribution should be considered to take into account the quantum nature of electrons with the statistical approach.

Under the quantization conditions, the total number of electrons in a volume V is given by

$$N = 2 \times \frac{1}{(2\pi)^3} \frac{4}{3} \pi k_F^3 V, \tag{2.8}$$

where k_F is Fermi momentum (called Fermi wave vector) and the factor 2 has been drawn from the electronic spin. Therefore, the Fermi momentum is found $k_F =^3 \sqrt{3\pi^2 n}$ when the density $n = N/V$. It means that the density of electron determines the average momentum of electrons in the ensemble and the kinetic energy of the free electrons, so-called *Fermi energy*, is given by

$$E_F = \frac{\hbar^2 k_F^2}{2m} = \frac{\hbar^2}{2m} (3\pi^2 n)^{2/3}. \tag{2.9}$$

The typical metallic electron densities are usually in the range of 10^{22}–$10^{23}\,\mathrm{cm}^{-3}$ and, by the substitution, the kinetic energy of electrons is $E_F \sim 1$–$20\,\mathrm{eV}$ in an atomic energy unit and $E_F/k_B \sim 100 \times 10^3\,\mathrm{K}$ ($\gg$ room temperature). The Fermi wave vector gives the scales for various properties of a solid

$$k_F \sim \frac{1}{\text{atom spacing}} \sim \text{size of Brillouin zone} \tag{2.10}$$

$$\sim \text{typical X-ray } k\text{-vector}. \tag{2.11}$$

Furthermore, electron velocity at k_F can be estimated as $v_F = \frac{\hbar k_F}{m} \sim 10^{-2} c$ which is quite fast even at a ground state $T = 0$. The value is 10 times faster than what Drude model predicted for. This implies an average distance traveled between collisions 10 times longer, suggesting that the collisions are due to something different.

From the quantization condition, the temperature dependence of electron's heat capacity can be driven. Electron density follows Fermi–Dirac statistics that the occupation of an energy state by an electron is either 0 and 1. Using the partition function $Z = \sum_i e^{-\beta E_i}$ which describes the sum of the probability distribution in the energy states E_i at a finite $(k_B T)^{-1} = \beta$, the Fermi–Dirac function reads

$$f = \frac{1}{e^{(E-\mu)/k_B T} + 1}, \tag{2.12}$$

where μ is the chemical potential. Here f gives the unnormalized probability of electron occupation at the energy E. The behavior of the function is sketched in Fig. 2.1. The Fermi energy E_F is $E_F = \mu$ at a zero temperature $T = 0$. At a low temperature $\mu \gg k_B T$, the energy distribution is spread around the chemical potential. Electrons with energies within $\sim k_B T$ of μ are able to contribute to thermal processes. Due to the Pauli exclusion principle, electrons with energy further below μ are not able to acquire enough thermal energy to be excited into the empty space. So only the electrons close to the Fermi energy will contribute to the electrical transport (conductivity).

Using the Fermi–Dirac distribution, it is possible to calculate internal energy in order to get the heat capacity. Internal energy $U(T)$ is the average energy of the electron gas at a finite temperature. From the Fermi

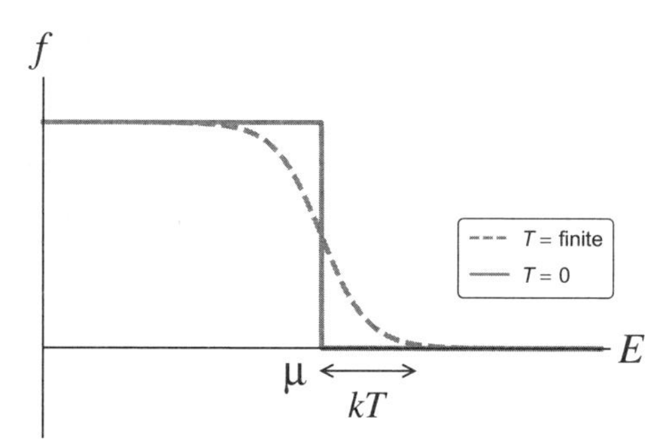

Fig. 2.1. The Fermi–Dirac distribution function at $T = 0$ and at a finite temperature T increased by $\mu \sim E_f$.

energy, the density function of the electrons becomes $D(E) \equiv dn/dE = \frac{1}{2\pi^2}\left(2m_e/\hbar^2\right)^{3/2} E^{1/2}$ and the internal energy is given by

$$U(T) = \int_0^\infty ED(E)f(E,T)dE \tag{2.13}$$

$$= \frac{1}{2\pi^2}\left(\frac{2m_e}{\hbar^2}\right)^{3/2}\int_0^\infty \frac{E^{3/2}dE}{e^{(E-\mu)/k_BT}+1} \tag{2.14}$$

$$\approx \frac{\mu^{5/2}}{5\pi^2}\left(\frac{2m_e}{\hbar^2}\right)^{3/2}\left[1+\frac{5}{8}\left(\frac{\pi k_BT}{\mu}\right)^2\right] \tag{2.15}$$

at a low temperature limit, where $\mu \gg k_BT$.[a] Therefore, the heat capacity of a solid is given by

$$C_{el} = \frac{dU(T)}{dT} \propto T. \tag{2.17}$$

The Sommerfeld model correctly predicts the heat capacity of a solid at a low temperature so that the theory is better than Drude's model. The model also produces good agreement to magnetic susceptibility which we will discuss in Section 2.6.

However, the model by Sommerfeld fails in explaining several other features in solid. For example, the free electron model cannot explain the differences between conductors, semiconductors and insulators. The model also cannot describe effectively the magnetic field dependence and magnetoresistance in Hall effect. The existence band structure arising from various shapes of Fermi surface cannot be explained in this model. Therefore, for the purpose of better understanding we need better model than Sommerfeld model.

The free electron model can be said as semiclassical model because the electron has been treated quantum mechanically. In fact, the electrons in a solid are not free but are usually bounded by the potential energy from the ions in the lattices.

[a] In the formula, we use the asymptotic form of Fermi–Dirac integrals,

$$F_a(x_0) = \int_0^\infty \frac{x^a}{e^{x-x_0}+1}dx \approx \frac{x_0^{a+1}}{a+1}\left(1+\frac{\pi a(a+1)}{6x_0^2}+O(x_0^{-4})\right). \tag{2.16}$$

2.3. Bloch's Theorem, Free Electrons and Tight-Binding Model

The problem with the free electron models is to ignore the interaction of electrons with the crystal lattice. The lattice itself is composed of periodically spaced ion cores that gives rise to a periodic potential variation for electrons moving through the crystal. In this section, we will discuss about how to deal with a periodic potential in a crystal which is known as Bloch theorem. From the formalism, it will be shown that the band structure of electron will emerge. The behavior of electrons in the periodic potential is dependent on the interaction strength and the case of weak potential as well as the tight-binding model will be discussed here.

2.3.1. *Bloch theorem: First quantization treatment*

Now we introduce the effect of periodic potential to the free electrons. The underlying translational periodicity of the crystal is defined by the lattice translational vectors

$$\bar{T} = n_1\bar{a}_1 + n_2\bar{a}_2 + n_3\bar{a}_3, \tag{2.18}$$

where n_i is integer numbers and $\bar{a}_i$ is primitive lattice vectors. The lattice translational vector describes the shape of ionic lattices in a crystal. Assuming that the potential energy $V(r)$ in the lattice is periodic with respect to spatial translation, it satisfies

$$V(\bar{r}) = V(\bar{r} + \bar{T}). \tag{2.19}$$

The periodic nature of the potential energy becomes easier to work with Fourier transformation

$$V(\bar{r}) = \sum_{\bar{G}} V_G \cdot e^{i\bar{G}\cdot\bar{r}}, \tag{2.20}$$

where $\bar{G}$ is a vector in the transformed space $\bar{G} = m_1\bar{A}_1 + m_2\bar{A}_2 + m_3\bar{A}_3$. Substituting (2.20) into (2.19), we obtain a condition $e^{i\bar{G}\cdot\bar{T}} = 1$ which gives $\bar{G}\cdot\bar{T} = 2\pi l$ with integer l. Then, it leads to the relationship between lattice vector $\bar{T}$ and $\bar{G}$ named reciprocal lattice

$$\bar{a}_j \cdot \bar{A}_k = 2\pi\delta_{jk}, \tag{2.21}$$

where $\bar{G}$ defines reciprocal lattice from the lattice vector. The reciprocal lattice also is periodic which implies the existence of lattice in a momentum space. The periodicity in momentum space implies that all information about the lattices lie within the primitive unit cell of the reciprocal lattice, known as a Brillouin zone.

We now inspect the Schrödinger equation for electrons in periodic potential. The Schrödinger equation

$$\left\{-\frac{\hbar^2}{2m}\nabla^2 + V(r)\right\}\psi(r) = E\psi(r) \tag{2.22}$$

has plane waves solution which has the form of series expansion in momentum space, $\psi(\bar{r}) = \sum_k C_{\bar{k}} e^{i\bar{k}\cdot\bar{r}}$. The periodicity of potential energy imposes periodic boundary conditions in the wave function,

$$e^{i\bar{k}(\bar{r}+N_j\bar{a}_j)} = e^{i\bar{k}\cdot\bar{r}} \tag{2.23}$$

and so that we have the condition $e^{iN_j\bar{k}\cdot\bar{a}_j} = 1$. Here N_j is the number of primitive unit cells in the jth direction. It is called Born–von Karman periodic boundary condition. Comparing the condition with reciprocal vector, the allowed wave vector for the wave function can be obtained

$$\bar{k} = \sum_{j=1}^{3} \frac{m_j}{N_j}\bar{A}_j. \tag{2.24}$$

This tells us that there are many k's where the state can occupy. As we changed the m_j, we generate a new state with different momentum. It is important consequences that the periodicity in the lattice is reflected in the periodic modulation of the electronic wave functions and the Brillouin zone always contains same number of momentum states as the primitive unit cell in the crystal.

The specified changes of wave function derived from the periodicity of the potential energy also can be evaluated by the Schrödinger's equation. Substituting the plan wave solution, the equation becomes

$$\sum_k e^{i\bar{k}\cdot\bar{r}}\left\{\left(\frac{\hbar^2k^2}{2m} - E\right)C_k + \sum_G V_G C_{k-G}\right\} = 0 \tag{2.25}$$

and, due to the orthogonal condition, we have

$$\left(\frac{\hbar^2k^2}{2m} - E\right)C_k + \sum_G V_G C_{k-G} = 0. \tag{2.26}$$

When we take $V_G = 0$, the state becomes equivalent to the Sommerfeld model without the potential energy. Thanks to the periodicity of the potential energy, we only need to solve the equation for the first Brillouin zone which is still not easy to solve in general. To take into account a general case, we write $\bar{k} = \bar{q} - \bar{G}'$ where $\bar{q}$ lies in the first Brillouin zone and $\bar{G}'$ is a reciprocal lattice vector. Then we get

$$\left(\frac{\hbar^2(\bar{q} - \bar{G}')^2}{2m} - E\right) C_{\bar{q}-\bar{G}'} + \sum_{G} V_G C_{\bar{q}-\bar{G}'-\bar{G}} = 0. \tag{2.27}$$

Finally, we change our variables so that $\bar{G}'' \to \bar{G} + \bar{G}'$. Then, with the periodic potential, the equation is invariance under the translation of reciprocal vector $\bar{G}'$. Choosing a particular value of $\bar{q}$, the only $C_{\bar{k}}$ featuring the equation above is of the form $C_{\bar{q}-\bar{G}}$ and it will return to the wave function,

$$\psi_{\bar{q}}(\bar{r}) = \sum_{\bar{G}} C_{\bar{q}-\bar{G}} e^{i(\bar{q}-\bar{G})\cdot\bar{r}} \tag{2.28}$$

$$= e^{i\bar{q}\cdot\bar{r}} \sum_{\bar{G}} C_{\bar{q}-\bar{G}} \cdot e^{-i\bar{G}\cdot\bar{r}} = e^{i\bar{q}\cdot\bar{r}} u_{\bar{q}}(\bar{r}), \tag{2.29}$$

where $u_{\bar{q}}(\bar{r})$ is invariant for the translation by lattice vector $u_{\bar{q}}(\bar{r}) = u_{\bar{q}}(\bar{r} + \bar{T})$. Therefore, the solution under the periodic potential is a plane wave which is modulated by a function with lattice periodicity. This is called *Bloch theorem*. This is quite interesting result which do not assume anything about the type of particle and the strength of potential except the periodicity of potential energy. The theorem can be applied to argue whether a material behaves as an insulator, semiconductor or metal. The issue will be handled in the later section.

Apart from the fact that the wave function is modulated by periodic potential, it is still not easy to find the solution for $u_{\bar{q}}$ in general. But we look at special cases in the following sections (Fig. 2.2).

2.3.2. *Weak potential — Nearly free electron model*

Considering the case when the potential energy in the Schrödinger equation is very small compared with their kinetic energy. This means that the electrons in the lattices are nearly free and the periodic contribution to the energy can be taken as a small perturbation. When the potential energy is nearly zero, the dispersion relation is given by free particle energy $E = \hbar^2 k^2/2m = \hbar^2(q - G)^2/2m$ which is periodic in the momentum space, reciprocal space, as sketched in Fig. 2.3.

In the nearly free electron limit, the Schrödinger equation returns the free electron dispersion relation at the momentum space except the points

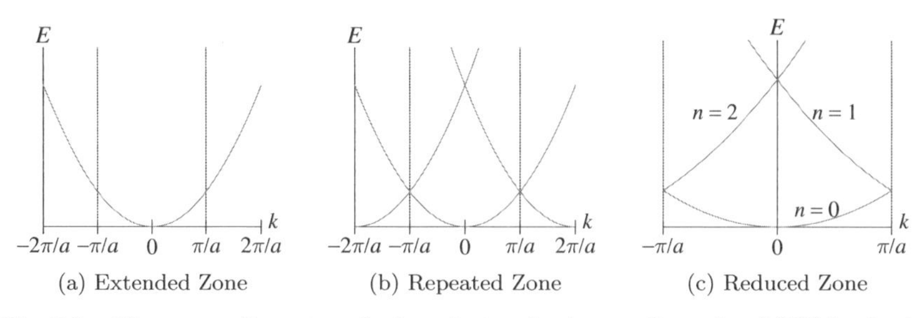

Fig. 2.2. The energy dispersion of a free electron in the one-dimensional (1D) lattice in three different ways. In the extended zone scheme, the energy E^0_{k-G} is considered as all possible k for $G = 0$ in the whole reciprocal space. In the repeated zone scheme, E^0_{k-G} is described for all G. Finally, the reduced zone scheme, the most widely used scheme, shows the energy only in the first Brillouin zone ($-\pi/a < k < \pi/a$) and newly labels the band index n originated in the index G.

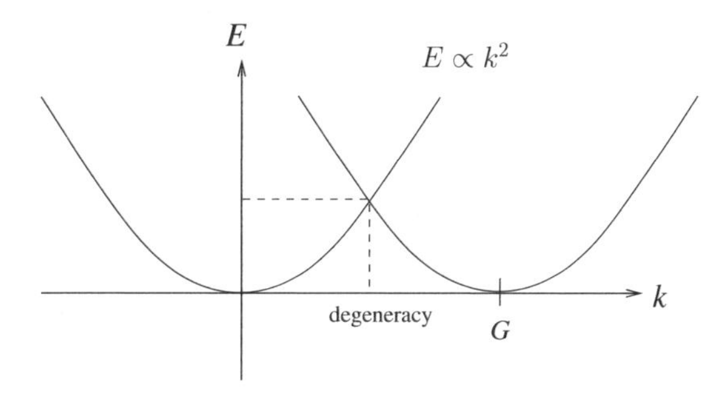

Fig. 2.3. Periodic electron dispersion relation when the potential energy is negligible.

where two curves are overlapped. This is the point where Bragg reflection occurs due to the interference of two wave functions. At the point q and $q - G$, the degeneracy exists, the term of perturbation is nonetheless trivial and the Schrödinger equation (2.27) returns two none trivial potential parts

$$\left[\frac{\hbar q^2}{2m} - E\right] C_q + V_G C_{q-G} = 0, \tag{2.30}$$

$$\left[\frac{\hbar (q-G)^2}{2m} - E\right] C_{q-G} + V_{-G} C_q = 0. \tag{2.31}$$

With the symmetric potential energy, $V_G = V_{-G}$, the linear equations can be solved and the energy eigenvalues can be obtained

$$E_\pm = \frac{1}{2}(E_q + E_{q-G}) \pm \sqrt{\frac{(E_q - E_{q-G})^2}{4} + V_G^2} \tag{2.32}$$

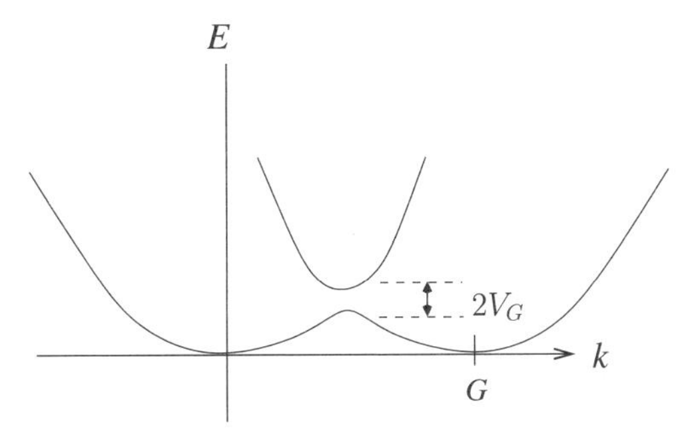

Fig. 2.4. The evolution of nearly free electron bands are opening up at the point on the boarder of Brillouin zone boundary.

with simplified notation $E_k = \hbar k^2/2m$. From the periodicity condition the kinetic energy of electron is $E_k \approx E_{k-G}$ and the energy eigenvalues are found as $E_{\pm} = \frac{1}{2}(E_k + E_{k-G}) \pm V_G$. At the point of degeneracy, band gap opens up $\Delta E = E_+ - E_- = 2V_G$. Because of continuity we have the band gap as can be seen as it is appeared in Fig. 2.4. This happens when $|k| \approx |k - G|$, i.e. when $\bar{k}$ lies in plane halfway between 0 and $\bar{G}$. Therefore,

$$\bar{G} \cdot \left(\bar{k} - \frac{1}{2}\bar{G} \right) = 0 \tag{2.33}$$

or $G^2 = 2\bar{k} \cdot \bar{G}$ (same as band gap). Generally speaking, the relation provides momentum profile at the point of band gap when the lattice vector has been specified.

2.3.3. *Tight-binding model*

In the tight-binding model, it is assumed that the strength of the crystal potential is large, which means that the ionic potentials are very strong. It follows, therefore, that when an electron is captured by ion during its motion through the lattice, the electron remains within the range of ionic potential for a long time before hopping to the next ion. It implies that the kinetic energy of the electrons is appreciably lower than the top of the potential barrier.

Assume that electrons are confined closely to atoms in the lattice n. For simplicity, imagine a single energy level E_n and then the electrons are tunneling to nearest neighbors with the strength A_n. Then, the Hamiltonian can be written only by the site energy and electronic hopping terms

$$H = \sum_n E_n|\phi_n\rangle\langle\phi_n| + A_n\left(|\phi_n\rangle\langle\phi_{n-1}| + \text{h.c.}\right), \tag{2.34}$$

where $|\phi_n\rangle$ is a wave function localized around nth site. The expressions for E_n and A_n are obtained from the Schrödinger equation:

$$E_n = \langle\phi_n|\left(-\frac{\hbar^2\nabla^2}{2m} + V(r)\right)|\phi_n\rangle, \tag{2.35}$$

$$A_n = \langle\phi_n|\left(-\frac{\hbar^2\nabla^2}{2m} + V(r)\right)|\phi_{n-1}\rangle. \tag{2.36}$$

The wave function $|\phi_n\rangle$ which represents the localized electrons at nth site is called Wannier function. The wave function can be as a function of the ion's position

$$\phi_n(r) = \langle r|\phi_n\rangle = g(r - r_n) \propto e^{-\frac{(r-r_n)^2}{2\sigma^2}}, \tag{2.37}$$

which is Gaussian function centered at r_n. To satisfy the tight-binding condition, the overlap of the wave function is non-trivial only between the neighbor sites so that the orthogonality condition of the function is provided as

$$\int g(r-r_n)g(r-r_{n\pm1})dr \neq 0, \quad \int g(r-r_n)g(r-r_{n\pm k\geq 2})dr = 0. \tag{2.38}$$

The conditions are originated from only counting the nearest neighbor interactions. This describes primarily the low-lying narrow bands for which the shell radius is much smaller than the lattice constant. The tight-binding model can be solved by evaluating the Hamiltonian (2.35). From the function, the coefficients in the Hamiltonian can be evaluated as

$$\begin{aligned}\langle\phi_n|&\left(-\frac{\hbar^2\nabla^2}{2m} + V(r)\right)|\phi_n\rangle \\ &= \int\int g(r-r_n)\left(-\frac{\hbar^2\nabla^2}{2m} + V(r)\right)g(r'-r_n)drdr'.\end{aligned}$$

In general, we can treat the problem in three spatial dimension by taking r as a 3D vector $\vec{r}$. In that case, the differentiation in the spherical coordinates becomes

$$\nabla^2 = \frac{1}{r^2 \sin\theta}\left[\sin\theta\frac{\partial}{\partial r}\left(r^2\frac{\partial}{\partial r}\right) + \frac{\partial}{\partial\theta}\left(\sin\theta\frac{\partial}{\partial\theta}\right) + \frac{1}{\sin\theta}\frac{\partial^2}{\partial\phi^2}\right].$$

We consider 1D lattice array only. To find the dispersion relation, energy spectrum of the Hamiltonian is need to be found. The Hamiltonian (2.35) can be diagonalize using Fourier transformations

$$|\psi_k\rangle = \sum_n e^{ikr_n}|\phi_n\rangle, \tag{2.39}$$

$$|\phi_n\rangle = \sum_k e^{-ikr_n}|\psi_k\rangle, \tag{2.40}$$

and if we insert the second equality into the Hamiltonian, we have

$$H = \sum_n \Bigg[E_n \sum_{kl} e^{-i(k-l)r_n}|\psi_k\rangle\langle\psi_l| \tag{2.41}$$

$$+A_n \sum_{kl} e^{-ikr_{n-1}}e^{ilr_n}|\psi_k\rangle\langle\psi_l| + \text{c.c.}\,|\psi_k\rangle\langle\psi_l|\Bigg] \tag{2.42}$$

where c.c. means complex conjugation. The periodicity of the lattice potential implies that the coefficients of the parameters are constant throughout the lattices. It means $E = E_n$, $A = A_n$ and $r_n = n \cdot a$ where a is lattice space and, with the conditions, the Hamiltonian is diagonalized as

$$H = E\sum_k |\psi_k\rangle\langle\psi_k| + A\sum_k \cos(ka)|\psi_k\rangle\langle\psi_k| \tag{2.43}$$

$$= \sum_k \left[E + 2A\cos(ka)\right]|\psi_k\rangle\langle\psi_k| \tag{2.44}$$

so that the energy dispersion relation with respect to the momentum is periodic as $E(k) = E + 2A\cos(ka)$. The above treatment can be extended to three dimensions in a straightforward manner. Thus, for the case of scattering lattices, the band energy is given by

$$E(k) = E + 2A[\cos(k_x a) + \cos(k_y a) + \cos(k_z a)] \tag{2.45}$$

when the lattice space constant a is homogeneous in every directions. Therefore, the energy contours for this band, in the any of paired plane, e.g. k_x–k_y plane, are periodic and symmetric under the homogenous condition. It means that the shape of the band in momentum space will be determined by the real-space crystal structure. If the atoms in a certain direction are far apart, the bandwidth will be narrow for the motion in that direction.

In fact, the behavior of electrons in the nearly free electron model and the tight-binding model is qualitatively quite similar. Both band structures have minima and maxima at the Brillouin-zone centers and boundaries, respectively. They also have same momentum space periodicity. More complex arrangements of atoms or molecules often found in real solids will give rise to more complex band shapes, but qualitatively the properties of the bands will be the same as those of the simple models.

In the formula, there is net momentum k appeared. However, one should be careful that it is not the real momentum of electron. It can be identify as, in the wave function of the Bloch theorem, $\hbar k$ is not the eigenvalues of free electron. k is the index for the Fourier transformation and it physically means *crystal momentum* which is effective momentum from the periodicity of electron. From the Bloch form of a state, the group velocity of the electrons

$$v = \frac{1}{\hbar}\nabla_k E \tag{2.46}$$

can be found, where ∇_k is the gradient operator in k-space, to describe the real space motion of electron. In this regards, effective mass of electron m^* is possible to be derived from the Newton equation. An acceleration a is related with an effective force F acting on an electron

$$a = \frac{dv}{dt} = \frac{dv}{dk}\frac{dk}{dt} = \frac{1}{\hbar^2}\frac{d^2E}{dk^2}F \tag{2.47}$$

so that the effective mass of electrons in the lattices is given by

$$m^* = \hbar^2 / \left(\frac{d^2E}{dk^2}\right). \tag{2.48}$$

Therefore, the dispersion relation of electrons can be used for the effective masses and the Bloch electron behaves like a free electron whose mass is m^*. As an example, the effective mass m^* in tight-binding model can be evaluated from (2.43). The dispersion relation for the model with its Taylor expansion is given by

$$E = 2A\cos ka + \text{const} \tag{2.49}$$

$$= 2A\left(1 - \frac{1}{2}k^2a^2 + \cdots\right) + \text{const.} \tag{2.50}$$

Therefore, in the small k region, one finds that the effective mass is inversely proportional to the hopping rate and the square of lattice constant as

$$\frac{d^2E}{dk^2} = -2A \cdot a^2, \quad m^* = -\frac{\hbar^2}{2Aa^2}. \tag{2.51}$$

Interestingly, the mass can be positive or negative depending on the sine of A. Dynamically, the negative mass can be seen as the negative acceleration whose direction is opposite to the force applied. This means that, in the region, the lattice exerts a retarded force on the electron that it overcomes the applied force and produces a negative acceleration. Physically, the positive mass implies the localization of electrons and the negative mass favors hopping of electrons in the lattices.

2.3.4. *Case study: Graphene*

In this subsection, we will try to investigate the band gap structure of graphene which has a specific electronic structure of an isolated carbon atom. For the case of graphene, the three electrons in the outer orbitals are arranged themselves in a planer shape at the 120° angles and the lattice is thus formed in the shape of the honeycomb lattice. It is notable that there are two inequivalent sublattices (A and B in its choices of notation) with the environments of the corresponding carbon atoms being mirror images of one another. The honeycomb 2D lattice is composed of the basic cell of the system which has the primitive lattice vectors for the atoms of same species as

$$\vec{a}_1 = \frac{a_0}{2}(3\hat{x} + \sqrt{3}\hat{y}), \quad \vec{a}_2 = \frac{a_0}{2}(3\hat{x} - \sqrt{3}\hat{y}), \tag{2.52}$$

where a_0 is the nearest neighbor bonding spacing as $a_0 = |\vec{a}_i|/\sqrt{3}$ whose values are given around 1.42 Å. The reciprocal vector can be obtained to satisfies the condition, $\vec{a}_i \cdot \vec{G}_j = 2\pi\delta_{ij}$ and the first Brillouin zone of the reciprocal lattice in the standard way. They are

$$\vec{G}_1 = \frac{2\pi}{3a_0}(\hat{x} + \sqrt{3}\hat{y}), \quad \vec{G}_2 = \frac{2\pi}{3a_0}(\hat{x} - \sqrt{3}\hat{y}). \tag{2.53}$$

The lattice structure of the graphene is shown in Fig. 2.5: there are A and B atoms in the unit cell with a reference point r_n and the position of A atoms is set as a reference point.

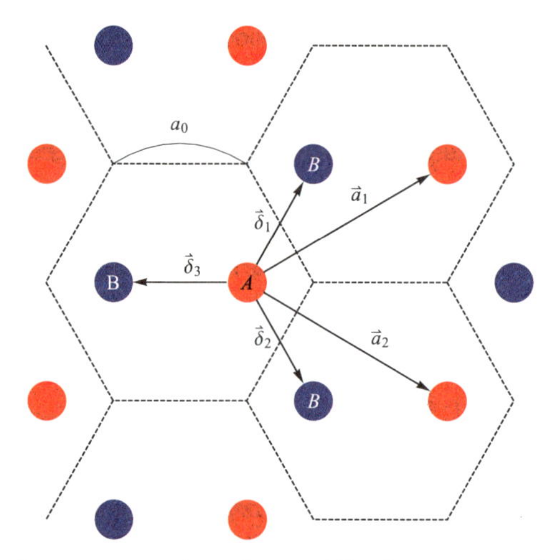

Fig. 2.5. The graphene structure in real space. The A atoms are red and B atoms blue. The dashed hexagon represents the unit cell of length $a_0 = 1.42$ Å.

For the first approach to the electronic band structure, it can be modeled by a tight-binding model with nearest neighbor hopping only. If we denote the orbital on atom i with spin σ by (i, σ) and the corresponding creation operator by $a^\dagger_{i\sigma}(b^\dagger_{i\sigma})$ for an atom on the two different sublattice, then the nearest neighbor tight-binding Hamiltonian has the simple form

$$\hat{H}_{tb,n.n.} = -t \sum_{i,j=n.n.\sigma} (a^\dagger_{i\sigma} b_{j\sigma} + \text{h.c.}), \tag{2.54}$$

where t is the electron hopping element. The numerical value of the hopping matrix element is believed to be about 2.8 eV. When the hexagonal sublattice A (or B) only composed of A (or B) atoms is considered, the wave function can be expressed by using the Wannier function $\phi_n(r)$ in Eq. (2.37)

$$\Phi_{A,\vec{k}}(\vec{r}) = \sum_n e^{i\vec{k}\cdot\vec{r}_n} \phi_{\vec{r}_n}(\vec{r}), \tag{2.55}$$

$$\Phi_{B,\vec{k}}(\vec{r}) = \sum_n e^{i\vec{k}\cdot(\vec{r}_n + \vec{\delta}_0)} \phi_{\vec{r}_n + \vec{\delta}_0}(\vec{r}), \tag{2.56}$$

where $\vec{\delta}_0$ is the nearest neighbor vector directing the B atom in the nth unit cell. Then, these wave functions Φ_A and Φ_B are the appropriate basis of the graphene. In the second quantization representation, the form of the eigenfunctions for the tight-binding model can be finally expressed as

$$\begin{pmatrix} \alpha_{\vec{k}} \\ \beta_{\vec{k}} \end{pmatrix} = \sum_n \exp[i\vec{k} \cdot \vec{r}_n] \begin{pmatrix} a_n \\ b_n e^{i\vec{k}\cdot\vec{\delta}_0} \end{pmatrix}, \tag{2.57}$$

in the nth unit cell. When we consider the total lattice structure, B atoms are located around an A atom. The Hamiltonian in the $\vec{k}$-space becomes

$$\hat{H}_{\vec{k}} = \begin{pmatrix} 0 & \Delta_{\vec{k}} \\ \Delta_{\vec{k}} & 0 \end{pmatrix}, \qquad \Delta_{\vec{k}} \equiv -t \sum_{l=1}^{3} \exp i\vec{k} \cdot \vec{\delta}_l, \tag{2.58}$$

where the three nearest neighbor vectors $\vec{\delta}_{l=1,2,3}$ are

$$\vec{\delta}_1 = \frac{a_0}{2}(\hat{x} + \sqrt{3}\hat{y}), \qquad \vec{\delta}_2 = \frac{a_0}{2}(\hat{x} - \sqrt{3}\hat{y}), \qquad \vec{\delta}_3 = -a_0\hat{x}. \tag{2.59}$$

By using these nearest neighbor vectors, we obtain

$$\Delta_{\vec{k}} = -t \exp(-ik_x a_0) \left(1 + 2 \exp\left(i\frac{3k_x a}{2} \right) \cos \frac{\sqrt{3}}{2} k_y a \right) \tag{2.60}$$

and the eigenvalues of $\hat{H}$, $\epsilon_{\vec{k}}$ are given by

$$\epsilon_{\vec{k}} = \pm|\Delta_{\vec{k}}| = \pm t \sqrt{1 + 4\cos\frac{3k_x a}{2} \cos\sqrt{3}\frac{k_y a}{2} + 4\cos^2\frac{\sqrt{3}}{2}k_y a}. \tag{2.61}$$

It is remarkable that (see Fig. 2.6) the minimum points of the upper band $\epsilon^+_{\vec{k}}$ meet the maximum points of the lower band $\epsilon^-_{\vec{k}}$, which correspond to the points K and K' of the first Brillouin zone in the reciprocal space.

2.4. Conductors, Semiconductors, Insulators

In Section 2.3, we have studied how the electrons behave when they are located in a periodic array of ions. Following the Fermi–Dirac statistics, it has been seen that only two electrons can occupy one momentum state and the value of momentum that an electron can take is multiple function of an integer number as was discussed in the Sommerfeld model. From that fact, we know that the electrons in a solid occupy the energy states sequentially and the electrons fill the energy states in the energy–momentum dispersion relation in a discrete manner. Moreover, as we discussed in Chapter 1, there is energy level which cannot be occupied by electrons, unallowed energy state (called energy band or energy gap), due to the periodic

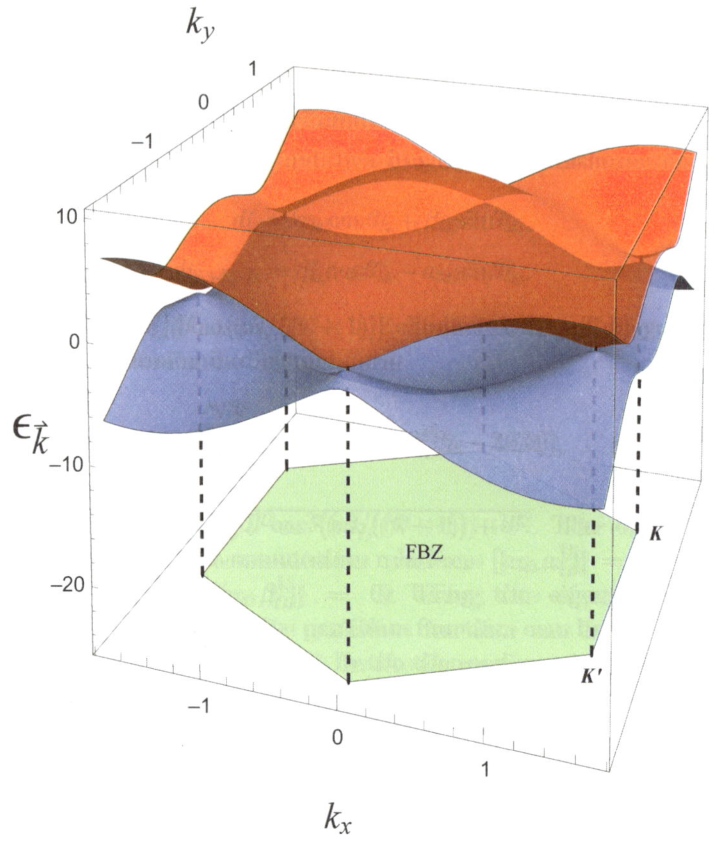

Fig. 2.6. The band structure of the graphene equation (2.61) in the reciprocal lattice space. The green hexagon describes the first Brillouin zone of the graphene.

structure of ions in the solid. Depending upon the shape of energy state occupation, solids are divided into two major classes: metals and insulators.

A metal (or conductor) is a solid in which an electric current flows under the application of an electric field. By contrast, application of an electric field sometimes produces no current if the solid is an insulator. There is a simple criterion for distinguishing between the two classes on the basis of the energy-band theory. This criterion rests on the following statement: A band which is completely full carries no electric current, even in the presence of an electric field. It follows therefore that a solid behaves as a metal only when some of the bands are partially occupied.

In an insulator, all the states of low-lying bands are completely filled. Because the Pauli principle forbids multiple occupation of states by electrons, when an electric field is applied, the electrons are locked in place. The only possibility of electronic movement is through the electron's hopping to an upper band. However, the energy gap $\mathcal{E}_g$ between the upper band and lower bands acts like a physical barrier and the motion of electrons can only occur through quantum mechanical tunneling, whose rate is roughly $\exp(-k_F\mathcal{E}_g/(eE))$, where E is the strength of the applied electric field and k_f is the Fermi waver vector; the formal expression of the tunneling probability can be found in [6, Chapter 16]. By contrast, in a metal, at least one band is partly filled. In that case, electrons at the Fermi surface are free to move into adjoining states. In the presence of an electric field, the distribution of electrons slides slightly in the direction of the field, in momentum space, giving the electron population bearing a net momentum which leads to current flow.

The behavior as a conductor can be illustrated by Na. For Na, since the inner bands 1s, 2s, 2p are all fully occupied, the electrons in the shell do not contribute to the current. The electrons related with the current are located at the topmost occupied band, i.e. the valence band. In Na, this is the 3s band. This bend can accommodate $2N_c$ electrons, where N_c is the total number of unit cells. Now in Na, a Bravais bcc lattice, each cell has one atom, which contributes one valance (or 3s) electron. Therefore, the total number of valence electrons is N_c, and as these electrons occupy the band, only half of it is filled, as shown in Fig. 2.7. Thus, sodium behaves like a metal because its valence band is only partially filled.

In a similar fashion, it can be concluded that the other alkalis, Li, K, etc., are also metals because their valence bands — the 2s, 4s, etc., respectively — are only partially full. The noble metals, Cu, Ag, Au, are likewise conductors for the same reason. Thus, in Cu the valence band (the 4s band) is only half full, because each cell in its fcc structure contributes only one valence electron.

As an example of a good insulator, diamond (carbon) can be mentioned. In the carbon, the top band originates from a hybridization of the 2s and 2p atomic states which gives rise to two bands split by an energy gap. Since these bands arise from s and p states, and since the unit cell here contains two atoms, each of these bands can accommodate $8N_c$ electrons. Now in diamond each atom contributes four electrons, resulting in eight valence electrons per cell. Thus, the valance band here is completely full and the substance is an insulator, as is stated above.

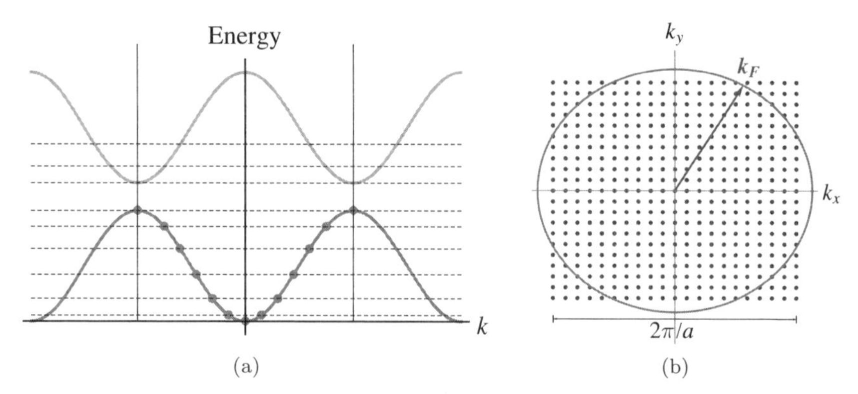

Fig. 2.7. The distribution of electrons in the diagram of dispersion relation. (a) 1D picture in the dispersion relation and (b) 2D picture within the Fermi surface when each lattice site has two conduction electrons.

There are additional classes of solids which can be defined using the concepts of filled bands. A *semiconductor* is an insulator where the energy gap $\mathcal{E}_g$ is small enough compared to $k_B T$ at room temperature that thermal fluctuations provide a substantial population of conducting electrons. Roughly speaking, the definition of a semiconductor is an insulator with a gap of 1–2 eV or less. Examples are Si and Ge, in which the gaps are about 1 and 0.7 eV, respectively. By contrast, the gap in diamond is about 7 eV. The conductivity of a typical semiconductor is very small compared to that of a metal, but it is still many orders of magnitude larger than that of an insulator. It is justifiable, therefore, to classify semiconductors as a new class of substance, although they are, strictly speaking, insulators at very low temperatures.

In some substances the gap vanishes entirely, or the two bands even overlap slightly, and we speak of *semimetals*. A semimetal is by contrast a metal with such a very small population of conduction electrons at zero temperature that its conducting properties are poor; a semimetal results when only a tiny pocket of electrons escapes the boundaries of the Brillouin zone, leading to conduction electron densities three or four orders of magnitude less than the normal metal. The best-known example is Bi, but other such substances are As, Sb, and white Sn.

A substance in which the number of valence electrons per unit cell is odd is necessarily a metal, since it takes an even number of electrons to fill a band completely. But when the number is even, the substance may be either an insulator or a metal, depending on whether the bands are disparate or overlapping.

2.5. Electron–Electron Interaction and Hartree–Fock Methods

Till now, we have considered the wave function of electrons that are confined in a periodic potential assuming that they are not interacting each other. In a realistic situation, the electrons tend to interact each other and the problem of practical band structure with the electron–electron interaction is normally taken as quite complicated problem.

Taking into account the electron–electron interactions together with the Born–Oppenheimer approximation,[b] the Schrödinger equation becomes

$$\hat{H}\psi = -\frac{\hbar^2}{2m}\sum_{l=1}^{N}\nabla_l^2\psi + \sum_{l=1}^{N}U_{\text{ion}}(\vec{r}_l)\psi + \sum_{l<l'}\frac{e^2}{|r_l - r_{l'}|}\psi = \mathcal{E}\psi, \qquad (2.62)$$

where the first term is the kinetic energy of the lth electron, the second term is the potential energy given by ions in the solid and the third term is the electron–electron interaction term. In the equation, we omit the position and the spin dependence of electronic wave function whose complete description should be $\psi = \psi(\vec{r}_1\sigma_1, \vec{r}_2\sigma_2, \ldots, \vec{r}_N\sigma_N)$ where $\vec{r}_i$ and σ_i are the position and the spin of an electron.

Considering all those terms in the Hamiltonian (2.62) and the symmetric conditions, the solution of the Hamiltonian is nonetheless trivial and should be treated with heavy computational methods. However, there are certain effective approximation which simplifies the electron–electron interaction terms. An effective Hartree–Fock method is an example of a general method called "mean field theory" which was originally designed for the atom with many electrons. The principle in the Hartree–Fock equation is same as that of the mean field theory. The method simplifies the electron–electron interaction by looking at the average effect on all $N-1$ electrons to a given single electron as

$$U_{ee}(r) \approx \int dr' \frac{e^2 \sum_j |\psi_j(r)|^2}{|r - r'|}, \qquad (2.63)$$

where $\sum_j |\psi_j(r)|^2 \equiv n(\vec{r})$ corresponds to the number density of electrons and $\psi_j(r)$ is the wave function of a single electron at a position r. Now, an

[b] The statement of Born–Oppenheiner approximation is that the electron's response to an ion's displacement is instantaneous such that the wave function of ions can be factorized out from the electronic wave function. It means that the ionic part in the Hamiltonian can be written separately.

electron sees only the ionic potential and the effective electronic potential such that the Schrödinger equation becomes

$$\sum_{l=1}^{N}\left[-\frac{\hbar^2}{2m}\nabla_l^2 + U_{\text{ion}}(\vec{r}_l) + U_{ee}(\vec{r}_l)\right]\psi = \mathcal{E}\psi. \tag{2.64}$$

Without any other constraint (and the spin degree of freedom), the eigenequation is reduced into the summation of the "single electron equations" by taking the product of single electron functions, $\psi = \psi_1(r_1)\psi_2(r_2)\psi_3(r_3)\cdots\psi_N(r_N)$.

To illustrate how the approximation works, let us consider the case of two interacting electrons. We assume that

$$H = \frac{1}{2}(p_1^2 + r_1^2) + \frac{1}{2}(p_2^2 + r_2^2) + \alpha\left[\frac{1}{\sqrt{2}}(r_1 - r_2)\right]^2, \tag{2.65}$$

where the Hamiltonian of two electrons contains interaction which is appeared in the last term. We assumed the normalized momentum $p_1 = i\partial/\partial r_1$. In this case, it is possible to obtain exact ground state using the coordination transformation. If we introduce new coordinates in terms of the center of mass and the relative position,

$$R = \frac{r_1 + r_2}{\sqrt{2}}, \quad r = \frac{r_1 - r_2}{\sqrt{2}}, \tag{2.66}$$

then the position part of the original Hamiltonian is to be transformed and the transformed Hamiltonian becomes

$$H = \frac{1}{2}(p_R^2 + R^2) + \frac{1}{2}\left[p_r^2 + (1 + 2\alpha)r^2\right], \tag{2.67}$$

where p_R and p_r are the momentum of the center of mass and the relative motion, respectively. The transformation decouples the Hamiltonian into two parts, which results in the solution with the ground state wave function $\psi_{\text{Ground}} \propto e^{-\frac{1}{2}R^2}e^{-\frac{1}{2}\sqrt{1+2\alpha}r^2}$ and the ground state energy function $E_{\text{Ground}} = \frac{3}{2}(1 + \sqrt{1+2\alpha})$. It exemplifies the transformation in which the electrons become independent and it allows to write the total state in the product of individual electronic wave function. Similarly, Hartree–Fock assumes the wave function of two electrons $\psi = \varphi(r_1)\varphi(r_2)$ which is consistent with the spin singlet state. Under the condition, the

Hartree–Fock equation becomes

$$E \cdot \varphi(r_1) = \frac{1}{2}(p_1^2 + r_1^2) + \alpha \int \varphi_2^* \left[\frac{1}{\sqrt{2}}(r_1 - r_2)\right]^2 \varphi_2 dr \tag{2.68}$$

$$= \frac{1}{2}\left[p_1^2 + (1+\alpha)r_1^2\right] + \frac{\alpha}{2}\int \varphi_2^* r_2^2 \varphi_2 dr. \tag{2.69}$$

In this case, the equation is easy to solve such that the wave function becomes

$$\varphi(r_1) \propto e^{-\sqrt{1+\alpha}r_1^2} \tag{2.70}$$

and $E = \frac{3}{2}\sqrt{1+\alpha}(\frac{3\alpha+1}{2\alpha+1})$. Here the zeroth-order energy is equal to $2E$.

The simplification of the Hamiltonian in (2.64) implies that the total wave function can be decomposed into the single electron wave functions. However, the solution of the Hamiltonian is more complicated than that of the Pauli exclusion principle. It requires that the wave function $\psi = \psi(\vec{r}_1\sigma_1, \vec{r}_2\sigma_2, \ldots, \vec{r}_N\sigma_N)$ is to be antisymmetrized for the particle exchange so that an extra phase should be added if the particle indices are swapped. The antisymmetrized wave function is constructed by *Slater determinant* of the orthogonal single particle wave functions. The antisymmetrized form of the wave function in terms of the single electron wave functions becomes

$$\psi(\vec{r}_1\sigma_1, \vec{r}_2\sigma_2, \ldots, \vec{r}_N\sigma_N)$$
$$= \frac{1}{\sqrt{N!}}\sum_s (-1)^s \psi_{s_1}(\vec{r}_1\sigma_1)\psi_{s_2}(\vec{r}_2\sigma_2)\cdots\psi_{s_N}(\vec{r}_N\sigma_N)$$
$$= \frac{1}{\sqrt{N!}}\begin{vmatrix} \psi_1(\vec{r}_1\sigma_1) & \psi_1(\vec{r}_2\sigma_2) & \cdots & \psi_1(\vec{r}_N\sigma_N) \\ \vdots & \vdots & & \vdots \\ \psi_N(\vec{r}_1\sigma_1) & \psi_N(\vec{r}_2\sigma_2) & \cdots & \psi_N(\vec{r}_N\sigma_N) \end{vmatrix},$$

where the sum is taken over all permutations s of $1, 2, \ldots, N$.

Considering the antisymmetrization condition, the expectation values of the Hamiltonian can be obtained as

$$\langle\psi|\hat{H}|\psi\rangle = \sum_i\sum_\sigma \int dr \left[\psi_i^*(r\sigma)\frac{-\hbar^2\Delta^2}{2m}\psi_i(r\sigma) + U_{\text{ion}}|\psi_i(r\sigma)|^2\right]$$
$$+ \sum_{\sigma_1\sigma_2}\int dr_1 dr_2 \frac{e^2}{|r_1 - r_2|^2}\sum_{i<j}\left[|\psi_i(r_1\sigma_1)|^2|\psi_2(r_2\sigma_2)|^2\right.$$
$$\left. - \psi_i^*(r_1\sigma_1)\psi_j^*(r_2\sigma_2)\psi_i(r_2\sigma_2)\psi_j(r_1\sigma_1)\right]. \tag{2.71}$$

In the formula, we omitted the vector description for the position r for the matter of simplicity. As it can be identified from the original Hamiltonian, the last term of the equation is originated from the antisymmetrization of the wave function. In the evaluation of the equation, there are additional constraints due to the normalization condition for the single electron wave function $\psi_i(r_j\sigma_j)$ as $\sum_\sigma \int dr\psi_i^*(r\sigma)\psi_i(r\sigma) = 1$.

Under the constraint, it is possible to find the minimum energy if we use the variational methods with the Lagrange multiplier $\mathcal{E}_i$. If we try to find the minimum of the energy expectation value with respect to the functional variation of ψ_i^*, the constraint equation becomes

$$\frac{\delta}{\delta\psi_i^*}\left[\langle\psi|\hat{H}|\psi\rangle - \sum_i \mathcal{E}_i \sum_\sigma \int dr\psi_i^*\psi_i\right] = 0. \quad (2.72)$$

We only consider the diagonal elements of the constraint since it is always possible to find the diagonal basis of the wave function in general. In that case, the constraint condition leads to the Hartree–Fock equation,

$$\mathcal{E}_i\psi_i(r\sigma) = \left[\frac{-\hbar^2\nabla^2}{2m} + U_{\text{ion}}(r)\right]\psi_i(r\sigma) + \psi_i(r\sigma)\sum_{\sigma'}\int dr'\frac{e^2\sum_j|\psi_j(r'\sigma')|^2}{|r-r'|}$$
$$+\sum_j \psi_j(r\sigma)\sum_{\sigma'}\int dr'\frac{e^2\psi_j^*(r'\sigma')\psi_i(r'\sigma')}{|r-r'|}. \quad (2.73)$$

The Hartree–Fock equation can be taken as effective Schrödinger equation for the ground state of electrons in a solid when the electron–electron interactions exist.

2.5.1. *Hartree–Fock for jellium*

The Hartree–Fock equations can be solved exactly for *jellium* which is for large number of electrons. The name is originated from the comment by Tonks and Langmuir (1929) concerning plasmas that "when the electrons oscillated, the positive ions behave like a rigid jelly with uniform density of positive charge". Jellium is just a collection of electrons, into which ions are introduced as a spatially uniform background to maintain overall charge neutrality. The Hartree–Fock equations are in this case solved by plane waves.

Electrons with the ions which are uniformly distributed with density n_e maintain overall neutrality in the medium. In that case, the wave function of electrons can be written separately from that of ions. In addition, the total wave function of the electrons can be decomposed into two parts, the spatial part and the spin part, as

$$|\psi(r,\sigma)\rangle_l = |\phi(r)\rangle_l |\chi(\sigma)\rangle_l, \tag{2.74}$$

where $|\phi(r)\rangle_l$ is the wave function for the spatial part of lth spin and $|\chi(\sigma)\rangle_l$ is for the spin part of the spin. The vector notation has been introduced which is related with the wave function as $\langle r,\sigma|\psi(r,\sigma)\rangle_l = \psi_l(\vec{r},\sigma)$.

After the decomposition, the Hartree–Fock equation in (2.73) can be written only by the spatial part of the wave function

$$E_l|\phi(\vec{r})\rangle_l = \left[-\frac{\hbar^2\nabla^2}{2m} - \frac{N}{V}\int d\vec{r}'\frac{e^2}{|\vec{r}-\vec{r}'|} + \int d\vec{r}'\sum_{j=1}^{N}\frac{e^2|\phi_j(\vec{r}')|^2}{|\vec{r}-\vec{r}'|}\right]|\phi(\vec{r})\rangle_l$$
$$-\sum_{j=1}^{N}\delta_{\chi_l,\chi_j}|\phi(\vec{r})\rangle_j\int d\vec{r}'\frac{e^2\phi_j(\vec{r}')^*\phi_l(\vec{r}')}{|\vec{r}-\vec{r}'|}, \tag{2.75}$$

where the right-hand side of the equation contains the terms for the kinetic energy of the electron, electron's interactions with ions, Coulomb interaction between the electrons and the exchange interaction of the electrons, respectively.

The Hartree–Fock equation for the jellium has the plane wave solution. Inserting the plan wave function, we have

$$\phi_l(\vec{r}) = \langle\vec{r}|\phi(\vec{r})\rangle_l = \frac{e^{i\vec{k}_l\cdot\vec{r}}}{\sqrt{V}}. \tag{2.76}$$

In that case, the interaction between the electron and the ions is canceled with the electron–electron interactions such that the relevant terms are the kinetic energy and the exchange interaction only. Through the evaluation of the exchange part, it is possible to obtain the ground state energy of the lth electron as

$$\mathcal{E}_l = \frac{\hbar^2k_l^2}{2m} - \frac{2e^2}{\pi}k_F F\left(\frac{k_l}{k_F}\right), \tag{2.77}$$

where k_F is Fermi momentum and the Lindhard dielectric function $F(x) = \frac{1}{4x}[(1-x^2)\ln\frac{1+x}{1-x} + 2x]$.

The solution has singularity at $k_l = k_F$ and the group velocity of electrons (identified by $\partial\mathcal{E}/\partial k$) become infinity near the Fermi surface. It is because of the effect that the Hartree–Fock approximation missed out. For the case of electrons far apart, the strength of the interaction falls off much faster than $1/|r - r'|$ due to the screening effects. If the distance between two electrons is large, their relative effect is screened since many electrons in between adjust their positions to hide the distant electrons from each other.

2.5.2. *Density functional theory and numerical methods*

In comparison to the Hartree–Fock equation which minimizes energy with respect to the wave function, the density functional theory states that the ground state is a unique functional of density such that the energy should be minimized by electron density ρ. The Hohenberg–Kohn theorem addresses that the functional that delivers the ground state energy of the systems, gives the lowest energy if and only if the input density is the true ground state density. This is nothing but the variational principle, however, this time with density and not wave function.

In the modern density functional theory, for example Kohn–Sham approach, the system is replaced by a fictitious noninteraction systems with same density as the real system. It is normal that the treatment of the electron density is usually given by numerical methods and they will be illustrated as in the following subsection.

2.5.3. *Numerical methods*

The methods which are dealing with the general many electron Hamiltonian are not perfectly accurate in comparison to the experimental data. It is partial because of the computational complexity of the proposed Hamiltonian and, on the other hand, because of the imperfection of the model in a real situation. So a lot of methods concerning the electronic properties actually aim to choose a set of basis functions that approximate the real solutions as closely as possible.

Here, we would like to mention three techniques:

- linear augmented plane wave (LAPW),
- orthogonal plane waves (OPW),
- linear combination of atomic orbital (LCAO).

Augmented plane waves (APWs): The main idea of the APWs is the superposition between the wave function of an electron in the atomic region and a small number of plane waves in the flat interstitial region. The wave functions between two regions are appropriately intersected at the boundary of a periodic potential. Before defining the APWs, a special periodic potential is introduced for simple calculation on the boundary condition, which is called the *muffin-tin potential*. The muffin-tin potential formally can be defined by

$$U(\vec{r}) = \begin{cases} V(|\vec{r}-\vec{R}|), & \text{when } |\vec{r}-\vec{R}| < r_{\rm mt}, \\ V(r_{\rm mt}) = 0, & \text{when } |\vec{r}-\vec{R}| > r_{\rm mt}, \end{cases} \tag{2.78}$$

where $r_{\rm mt}$ is less than the half of the nearest neighbor distance. The APWs are defined as follows:

(1) $\phi_{E\vec{k}} = e^{i\vec{k}\cdot\vec{r}}$ in the interstitial region;
(2) in the muffin-tin region at $\vec{R}$, $\phi_{E\vec{k}}$ does satisfy the atomic Schrödinger equation

$$-\frac{\hbar^2}{2m}\nabla^2\phi_{E\vec{k}}(\vec{r}) + V(\vec{r})\phi_{E\vec{k}}(\vec{r}) = E\phi_{E\vec{k}}(\vec{r}); \tag{2.79}$$

(3) $\phi_{E\vec{k}}$ is continuous at the boundary between atomic and interstitial regions.

The potential $V(\vec{r}-\vec{R})$ in the muffin-tin region is spherically symmetric. $\vec{k}$ in the potential area is determined by the boundary and is influenced by the plane wave satisfying the relation $\hat{H}\phi_{\vec{k}} = \frac{\hbar^2k^2}{2m}\phi_{\vec{k}}$ in the interstitial region. It is remarkable that contrary to the continuity of $\phi_{E\vec{k}}$, the derivative of $\phi_{E\vec{k}}$ $\nabla\phi_{E\vec{k}}$ is discontinuous at the boundary between atomic and interstitial region.

Let us specifically find the form of the wave function by using the APW method and the muffin-tin potential. In the muffin-tin potential region ($r < r_{mt}$), the radial equation of (2.79) with the spherically symmetric potential can be written as

$$\frac{-\hbar^2}{2mr^2}\frac{\partial}{\partial r}r^2\frac{\partial}{\partial r}R_{lE}(r) + \left[V(r) + \frac{\hbar^2 l(l+1)}{2mr^2}\right]R_{lE}(r) = ER_{lE}(r). \tag{2.80}$$

Outside the interstitial region, there is no potential. Thus, the wave functions in both the regions can be expressed as

$$\psi_{E\vec{k}}(\vec{r}) = \begin{cases} \sum_{l=0}^{\infty}\sum_{m=-l}^{l} A_{lm}R_{lE}(r)Y_{lm}(\theta,\varphi), & r < r_{mt}, \\ 4\pi\sum_{l=0}^{\infty}\sum_{m=-l}^{l} i^l j_l(kr)Y^*_{lm}(\theta_k,\varphi_k)Y_{lm}(\theta,\varphi), & r > r_{mt}, \end{cases} \tag{2.81}$$

where A_{lm} is the coefficient, j_l the lth spherical Bessel function, $Y_{lm}(\hat{r})$ the spherical harmonic function, and $R_l(r)$ the radial wave function. Actually, Eq. (2.80) has two independent solutions. One has the divergence at the origin and the other has the divergence at the infinity. Due to the boundary of the muffin-tin potential, the divergence at the infinity is no more disadvantage to describe the wave function in APW method, which is the first one in Eq. (2.81). The second wave function is the plane waves and an expression in terms of the spherical harmonics. By using the continuity at the boundary of the muffin-tin potential $r = r_{MT}$, the wave function becomes finally

$$\phi_{E\vec{k}} = 4\pi \sum_{l=0}^{\infty} \sum_{m=-l}^{l} \frac{i^l j_l(kr_{mt}) Y_{lm}^*(\theta_k, \varphi_k)}{R_{lE}(r_{mt})} Y_{lm}(\theta, \varphi) R_{lE}(r) \tag{2.82}$$

for all E, $\vec{k}$, and $\vec{r}$.

From the definition of APW, the augmented plane waves approximate the solutions satisfying not the crystal Schödinger equation but also the atomic Schödinger equation. Hence, we should correct the solution to satisfy the crystal Schödinger equation by a superposition of APW's. For any reciprocal lattice vector $\vec{K}$, the APW $\phi_{E,\vec{k}+\vec{K}}$ satisfies the Bloch condition with wave vector $\vec{k}$, and therefore the expansion of $\psi_{\vec{k}}(\vec{r})$ will be of the form

$$\psi_{\vec{k}} = \sum_{\vec{G}} b_{\vec{k}+\vec{G}} \phi_{E,\vec{k}+\vec{G}}. \tag{2.83}$$

The coefficients $b_{\vec{k}+\vec{G}}$ of Eq. (2.83) and the energies can be obtained by using the variational principle in terms of $b^*_{\vec{k}+\vec{G}}$. Inserting Eq. (2.83) into $\langle\psi|\hat{H} - E|\psi\rangle = 0$, we have

$$\sum_{\vec{K},\vec{K}'} b^*_{\vec{k}+\vec{K}'} b_{\vec{k}+\vec{K}} \langle\phi_{\vec{k}+\vec{K}'}|\hat{H} - E|\phi_{\vec{k}+\vec{K}}\rangle = 0. \tag{2.84}$$

By taking the functional derivative in terms of $b^*_{\vec{k}+\vec{K}'}$, we have

$$\sum_{\vec{K}} \langle\phi_{\vec{q}}|\hat{H} - E|\phi_{\vec{k}+\vec{K}}\rangle b_{\vec{k}+\vec{K}} = 0, \tag{2.85}$$

with $\vec{q} = \vec{k} + \vec{K}'$ and $\vec{K} - \vec{K}' \to \vec{K}$. By calculating the elements $\langle\phi_{\vec{q}'}|\hat{H} - E|\phi_{\vec{q}+\vec{K}}\rangle$, one can construct the matrix and the eigenvalue problem and the energy E and the coefficient $b_{\vec{k}+\vec{G}}$.

Orthogonalized plane waves (OPWs): The main idea of orthogonalized plane waves is a combination of rapid oscillations in the core region and plane wave-like behavior out of the core region: the core wave functions and the valence electrons. The core wave functions are well localized about the lattice sites, which are mainly derived from the Wannier function or the LCAOs, and the wave functions of the valence electrons found with the remarkable probability in the region between the sites and will be approximated by a very small number of plane waves.

Let us designate the core states as the localized orbitals such that

$$|\Phi_{n\vec{k}}\rangle = \frac{1}{\sqrt{N}}\sum_{\vec{R}} e^{i\vec{k}\cdot\vec{R}}|w_{n\vec{R}}\rangle, \tag{2.86}$$

where n means the core energy levels at sites $\vec{R}$ and $w_n(\vec{r}-\vec{R}) = \langle\vec{r}|w_{n\vec{R}}\rangle$ is the Wannier function. By using the projection operator $(\mathbb{1}-P)$ with $P = \sum_n |\Phi_{n\vec{k}}\rangle\langle\Phi_{n\vec{k}}|$, the valence electron states are constructed by the plane waves orthogonal to the core state

$$|\Phi^v_{\vec{k}}\rangle \equiv (\mathbb{1}-P)|\vec{k}\rangle = |\vec{k}\rangle - \sum_n |\Phi_{n\vec{k}}\rangle\langle\Phi_{n\vec{k}}|\vec{k}\rangle, \tag{2.87}$$

where the superscript v means the valence electrons. By taking $\langle\vec{r}|$ on the both sides of Eq. (2.87),

$$\begin{aligned}
\Phi^v_{\vec{k}}(\vec{r}) \equiv \langle\vec{r}|\Phi^v_{\vec{k}}\rangle &= \langle\vec{r}|\vec{k}\rangle - \sum_n \langle\vec{r}|\Phi_{n\vec{k}}\rangle\langle\Phi_{n\vec{k}}|\vec{k}\rangle \\
&= \frac{1}{\sqrt{V}}e^{i\vec{k}\cdot\vec{r}} - \sum_n \left(\int d\vec{r}\,\langle\Phi^c_{n\vec{k}}|\vec{r}\rangle\langle\vec{r}|\vec{k}\rangle\right)\langle\vec{r}|\Phi^c_{n\vec{k}}\rangle \\
&= \frac{1}{\sqrt{V}}e^{i\vec{k}\cdot\vec{r}} + \sum_n \left(-\frac{1}{\sqrt{V}}\int d\vec{r}\,\Phi^{c\,*}_{n\vec{k}}(\vec{r})\,e^{i\vec{k}\cdot\vec{r}}\right)\Phi^c_{n\vec{k}}(\vec{r}).
\end{aligned} \tag{2.88}$$

Since the core wave functions $\Psi^c_{n\vec{k}}(\vec{r})$ are localized at each sites, the last term in Eq. (2.88) is very small between the sites, which is equivalent to the original idea that the plane waves are prevalent outside the core region. It is obvious that the valence wave function $\Psi^v_{\vec{k}}(\vec{r})$ is orthogonal to all the core wave functions and satisfies the Bloch condition. Therefore, we may seek an expansion of the actual electronic eigenstates of the Schrödinger equation as linear combinations of OPW's,

$$\psi_{\vec{k}}^{v} = \sum_{\vec{K}} c_{\vec{K}} \Phi_{\vec{k}+\vec{K}}^{v} \tag{2.89}$$

of which the coefficients $c_{\vec{K}}$ and the energies $E_{\vec{k}}$ can be determined by the variational principle.

OPW methods are often used when one carries out a first principle calculation numerically. As expressing matrix elements of an atomic potential in terms of OPW and calculating with large eigenvalue problems, one can get a good convergent values about electronic density and energies. Also, this method offers the rational reason why the nearly free electron approximation does work well in predicting the band structures of a variety of metals.

Linear combination of atomic orbitals (LCAOs): Most electrons in a crystal are located in closed shells, and the rest of them are located in the outermost shells so their amplitudes decay drastically away from the nucleus (remind that the wave function for hydrogen atom decays exponentially). LCAOs method is adequate for electrons of atoms right beneath the Fermi surface, but not for the conduction electrons. The main idea of the LCAO is that, first of all, a collection of atoms is considered to be isolated. Each atoms has their own energy levels $\phi_n^{\rm at}$ that electrons can occupy. Then, let these atoms gather atoms together as closely as possible. Most wave functions for inner electrons behave like the one for the isolated atoms. The outermost electrons farther than innermost electrons from the core in the atom, however, are apt to overlap the wave function for electrons in other atoms. Here, we describe this physical situation with the formulas of quantum mechanics.

Let $\phi_{n'}^{\rm at}$ be the wave function for an electron occupying an isolated atom. The index n' indicates an electron orbital. The wave function satisfies

$$\hat{H}^{\rm at}\phi_{n'}^{\rm at}(\vec{r}) = -\frac{\hbar^2}{2m}\nabla^2\phi_{n'}^{\rm at}(\vec{r}) + V^{\rm at}(\vec{r})\phi_{n'}^{\rm at}(\vec{r}) = E_{n'}^{\rm at}\phi_{n'}^{\rm at}(\vec{r}), \tag{2.90}$$

which is the Schrödinger equation for an isolated atom. Let us now get all isolated atoms together and form a crystal with the Bravais lattice vectors $\vec{R}$. The Hamiltonian is

$$\hat{H} = -\frac{\hbar^2}{2m}\nabla^2 + V(\vec{r}) = -\frac{\hbar^2}{2m}\nabla^2 + \sum_{\vec{R}} V^{\rm at}(\vec{r} - \vec{R}). \tag{2.91}$$

To find the appropriate a set of the basis wave functions for the crystal Hamiltonian equation (2.91), the atomic wave function $\phi_{n'}^{\rm at}$ is used such that

$$\Phi_{n'\vec{k}}(\vec{r}) = \frac{1}{\sqrt{N}}\sum_{\vec{R}} e^{i\vec{k}\cdot\vec{R}}\phi_{n'}^{\rm at}(\vec{r} - \vec{R}), \tag{2.92}$$

which satisfies the Bloch's theorem. Still, $\Phi_{n'\vec{k}}$ is not enough to be a eigenfunction of the Hamiltonian equation (2.91) because it is neither normalized, nor eigenfunctions of the Hamiltonian. Although all eigenfunctions of the form Φ are not the solution of the Hamiltonian equation (2.91), a linear combination of some wave functions $\Phi_{n',\vec{k}}$ is a good candidate for the solution of the crystal Schrödinger equation (2.91). Then, the solutions are of the form:

$$\psi_{n\vec{k}}(\vec{r}) = \sum_{n'} C_{nn'}\Phi_{n'\vec{k}}(\vec{r}). \tag{2.93}$$

Like the previous methods did, the coefficient $C_{nn'}$ and the energy E can be found by using the variational method.

$$\langle\psi_{n\vec{k}}|\hat{H} - E|\psi_{n\vec{k}}\rangle = \sum_{i,j} C_{ni}^* C_{nj}\langle\Phi_{i\vec{k}}|\hat{H} - E|\Phi_{j\vec{k}}\rangle = 0,$$

$$\frac{\delta}{\delta C^*}\sum_{i,j} C_{ni}^* C_{nj}\langle\Phi_{i\vec{k}}|\hat{H} - E|\Phi_{j\vec{k}}\rangle = 0,$$

$$\sum_j C_{nj}(H_{ij} - E_n S_{ij}) = 0, \tag{2.94}$$

where

$$H_{ij} \equiv \langle\Phi_{i\vec{k}}|\hat{H}|\Phi_{j\vec{k}}\rangle \quad \text{and} \quad S_{ij} \equiv \langle\Phi_{i\vec{k}}|\Phi_{j\vec{k}}\rangle. \tag{2.95}$$

From specific calculation of the energy $E_n = H_{ij}/S_{ij}$ from Eq. (2.94), the tight-binding model can be derived. Before calculating the energy, we assume that the interaction between the nearest sites is only considered and the only s-orbital which has the spherical symmetry is specialized. Then, one can concisely evaluate the S_{ss} and H_{ss} in the following:

$$
\begin{aligned}
S_{ss} = \langle \Phi_{s\vec{k}} | \Phi_{s\vec{k}} \rangle &= \sum_{\vec{R}\vec{R}'} \frac{e^{i\vec{k}\cdot(\vec{R}-\vec{R}')}}{N} \int d\vec{r}\, \phi_s^{\rm at}(\vec{r}-\vec{R})\phi_s^{\rm at}(\vec{r}-\vec{R}') \\
&= 1 + \sum_{\vec{a}} e^{i\vec{k}\cdot\vec{a}} \int d\vec{r}\, \phi_s^{\rm at}(\vec{r})\phi_s^{\rm at}(\vec{r}+\vec{a}) \\
&= 1 + \sum_{\vec{a}} e^{i\vec{k}\cdot\vec{a}} \alpha,
\end{aligned} \tag{2.96}
$$

where the vector $\vec{a}$ is the difference from any lattice point to its nearest neighbors and

$$
\alpha \equiv \int d\vec{r}\, \phi_s^{\rm at}(\vec{r})\phi_s^{\rm at}(\vec{r}+\vec{a}). \tag{2.97}
$$

In the same way,

$$
\begin{aligned}
H_{ss} &= \sum_{\vec{R}\vec{R}'} \frac{e^{i\vec{k}\cdot(\vec{R}-\vec{R}')}}{N} \int d\vec{r}\, \phi_s^{\rm at}(\vec{r}-\vec{R}) \left[-\frac{\hbar^2}{2m}\nabla^2 + V(\vec{r}) \right] \phi_s^{\rm at}(\vec{r}-\vec{R}') \\
&= E_s^{\rm at} \left(1 + \sum_{\vec{a}} e^{i\vec{k}\cdot\vec{a}} \alpha \right) + U + \sum_{\vec{a}} e^{i\vec{k}\cdot\vec{a}} t,
\end{aligned} \tag{2.98}
$$

where

$$
U \equiv \int d\vec{r}\, \phi_s^{\rm at}(\vec{r})[V(\vec{r}) - V^{\rm at}(\vec{r})]\phi_s^{\rm at}(\vec{r}), \tag{2.99}
$$

$$
t \equiv \int d\vec{r}\, \phi_s^{\rm at}(\vec{r})[V(\vec{r}) - V^{\rm at}(\vec{r}+\vec{a})]\phi_s^{\rm at}(\vec{r}+\vec{a}). \tag{2.100}
$$

Then, the energy of the crystal Hamiltonian by using the LCAO method becomes

$$
\begin{aligned}
E = \frac{H_{ss}}{S_{ss}} &= \frac{E_s^{\rm at}\left(1 + \sum_{\vec{a}} e^{i\vec{k}\cdot\vec{a}}\right) + U + \sum_{\vec{a}} e^{i\vec{k}\cdot\vec{a}}\, t}{1 + \sum_{\vec{a}} e^{i\vec{k}\cdot\vec{a}} \alpha} \\
&= E_s^{\rm at} + \frac{U + \sum_{\vec{a}} t\, e^{i\vec{k}\cdot\vec{a}}}{1 + \sum_{\vec{a}} \alpha\, e^{i\vec{k}\cdot\vec{a}}}.
\end{aligned} \tag{2.101}
$$

Note that the s-orbital wave function is spherical, thereby α satisfies, i.e. $\alpha(-\vec{a}) = \alpha(\vec{a})$. Also, $V(\vec{r}) - V^{\rm at}(\vec{r}+\vec{a})$ has the inversion symmetry from the

Bravais lattice structure, which implies the relation $t(\vec{a}) = t(-\vec{a})$. When α is small and negligible, the s-orbital energy is finally simplified as

$$E_{\vec{k}} \approx E_s^{\text{at}} + U + \sum_{\vec{a}} t \cos(\vec{k} \cdot \vec{a}), \tag{2.102}$$

which depends on the vector $\vec{k}$. Equation (2.102) consists of the energy of original atomic orbital, a constant correction due to interactions, and a hopping term proportional to t and depending on $\vec{k}$. This last term describes the electrons hopping from one atomic site to neighboring sites. The reason that the last term is called "hopping" is that the group velocity of the electrons is given as

$$\vec{v}_{\vec{k}} = \frac{1}{\hbar} \vec{\nabla}_{\vec{k}} E_{\vec{k}} = \frac{i}{\hbar} \sum_{\vec{a}} t \, \vec{a} \, e^{i\vec{k}\cdot\vec{\delta}}. \tag{2.103}$$

2.6. Problems

1. **Calculating off-resonant Rabi flopping:** Imagine that one electron is trapped in a double-well potential structure (the so-called *double quantum dot*). Suppose that the electron can occupy one energy level only within each well. Let the energy difference between these levels be E and let the hopping energy between the wells be J. Calculate the state of the electron at any later time t, given that at $t = 0$ electron is localized in one of the wells.
2. **One-dimensional weak potential:** Electrons of mass m are confined to one dimension. A weak periodic potential described by $V(x) = V_0 + V_1 \cos(2x/a) + V_2 \cos(4x/a) + \cdots$ is applied.
   (a) Under what conditions will the nearly free electron approximation work? Assuming that this is satisfied sketch the three lowest bands in the first Brillouin zone.
   (b) Calculate to first order the energy gap at $k = \pi/a$ (between first and second band) and $k = 0$ (between second and third bond).
3. **Two-dimensional weak potential — Insulator vs. conductor:** An electron of mass m is moving in a square lattice of spacing a. Use the nearly free approximation.
   (a) Assume one electron per site. Draw the Fermi surface in two dimensions. Is this a metal or insulator?
   (b) Assume two electrons per site. Draw the Fermi surface in two dimensions. Is this a metal or insulator?

4. **Square lattice:** Consider 2D electrons subject to a weak periodic potential on a square lattice of spacing $a = 5\,\text{Å}$. For k vectors away from the edges of Brillouin zone, the wave function is well described by plane waves (as we argued in the lectures). Assume that, in general, wave functions have the Bloch form (since the potential is periodic). Consider states with energy E and wave vector $k = 0.5\,\text{Å}^{-1}$ in the x direction (the y component of k is zero). What will be the three lowest energies at this wave number? What are the corresponding Bloch functions?
5. **Effective mass:** Calculate the effective mass tensor for electrons in a simple cubic tight-binding band at the center ($k = (0, 0, 0)$), at the face center ($k = (0, 1, 1)$) and at the corner of the Brillouin zone ($k = (1, 1, 1)$). What is the usefulness of the concept of effective mass?
6. **Debye frequency in three dimensions:** A monatomic cubic material has lattice spacing a. The longitudinal and transverse sound speed is approximately equal (label it by c). What is the Debye frequency?

Chapter 3

Theory of Magnetism and Many-Body Quantum Mechanics

Magnetic property of a material is obtained by the collective effect of each individual constituent and external parameters. The effect is manifested from the quantity of atoms or electrons whose combinations are dependent upon the structural property of the system. Together with electric property, it can be found that the magnetism is typical many-body effect of the provided systems and the understanding of magnetism is possible only after the complete analysis of whole structure of the material.

In this chapter, we start our discussion from the second quantized picture of quantum particles and their many-body effect on its wave function. Subsequently, we demonstrate how the exotic phase of quantum material, i.e. superfluity, superconductivity and so on, is possible through the derivation of particle interaction models and their solutions.

3.1. Fock Space, Fermion, Bosons

Quantum state is described by wave function which is used to produce an expectation value of a given physical quantity. The description is designed to satisfy the particle's statistics derived from the Heisenberg uncertainty relation which states that system's conjugate variables cannot be measured simultaneously with arbitrary precision. Such properties are well satisfied even when the quantum state is written in a complex vector space. The particle statistics is reflected in its operator algebra when the quantum particle is described by the second quantization formalism.

The second quantization formalism is based on the particle description of a quantum state in comparison to the wave function representation. They are equivalent expression in its results although the mathematical object that they are dealing with are not completely identical. In this section, we introduce the second quantization description of quantum state especially when the state is composed of more than a single particle. The main question in the section is how to treat many particles in a complex vector space.

If we start with time-dependent single particle Schrödinger equation, the wave function is the solution of a linear equation

$$\hat{H}\psi(x,t) = i\hbar\frac{\partial}{\partial t}\psi(x,t) \tag{3.1}$$

as is discussed in the earlier chapter. Same equation satisfies even for two particles by replacing the single particle wave function $\psi(x,t)$ by two particle wave function $\psi_{12}(x_1,x_2,t)$ which describes particles at the position x_1 and x_2 at a given time t. Without any other constraints and the interactions between the particles, the linearity of the Hamiltonian requires that the many-body Hamiltonian is the superposition of the individual systems $\hat{H} = \hat{H}_1 + \hat{H}_2 + \cdots$ and the wave function is to be factorized as $\psi_{12}(x_1,x_2,t) = \psi_1(x_1,t)\psi_2(x_2,t)$ for the particles located at x_1 and x_2.

Many particle quantum state satisfies additional conditions due to the indistinguishability. It means that quantum particles are completely identical as far as they are same kind of species. The fact is need to be reflected in the particle statistics. The symmetry need to be dealt with care in the case of the many-body wave function. Depending on the particle's nature, exchange of two particles preserves the total wave function or generate extra phase -1 in the total wave function. The symmetrized and the antisymmetrized wave functions under the particle exchange are distinguished in the two different particles as

- Fermions: $\psi_{12}(x_1,x_2) = \psi_1(x_1)\psi_2(x_2) - \psi_2(x_1)\psi_1(x_2)$,
- Bosons: $\psi_{12}(x_1,x_2) = \psi_1(x_1)\psi_2(x_2) + \psi_2(x_1)\psi_1(x_2)$,

where the time dependence of the wave function is omitted in the equation. From the (anti) symmetrization of wave function, Fermions earn extra phase -1 under the particle exchange x_1 and x_2 while the wave function of Boson is remained same.

As it turn out, it is much more convenient to think of the wave function ψ as an operator $\hat{\psi}$. In the notion of second quantization,

$\hat{\psi}(x,t)^{\dagger}$[a] means the creation of a particle at position x and time t. Then, the particle statistics of Fermion and Boson are encoded in the commutation relation of the operators as

$$\begin{aligned} [\hat{\psi}(x), \hat{\psi}(y)^{\dagger}] &= \delta(x-y) \quad \text{Boson}, \\ \{\hat{\psi}(x), \hat{\psi}(y)^{\dagger}\} &= \delta(x-y) \quad \text{Fermions}, \\ \{\hat{\psi}(x), \hat{\psi}(y)\} &= [\hat{\psi}(x), \hat{\psi}(y)] = 0 \quad \text{Boson and Fermions}, \end{aligned} \tag{3.2}$$

where the square bracket denotes anticommutation relation of the operators, i.e. $[A, B] \equiv AB - BA$, and the curly bracket is that of commutation relation, $\{A, B\} \equiv AB + BA$.

To specify the exact meaning of the particle operator $\hat{\psi}(x)$, we need to define number states of a particle, called Fock state. Fock state denotes the particle's number in a vector space when it is generated from a vacuum state by adding arbitrary number of particles, say n, having specific momentum, say the kth mode. A vector representation of such a state for bosonic/fermionic particles is

$$|n\rangle_k = \frac{(a_k^{\dagger})^n}{\sqrt{n!}}|0\rangle_k, \tag{3.3}$$

where $a_k^{\dagger}$ is the creation operator of a bosonic/fermionic particle in the kth mode. The vector $|0\rangle_k$ denotes a vacuum state of k-mode meaning that there no particle which bears the momentum k. Due to the particle statistics, the structure of the fock space is changed dependent upon the type of particle. For a bosonic particle, the particle number n in $|n\rangle$ can be arbitrary positive number while for fermion n can take either value 1 or -1. We would discuss the structure of fock space in the following subsection.

3.1.1. *Second quantization for boson: Phonon as its example*

When the particle is boson, the application of the creation operators n times to the vacuum state generates n particles in the kth mode, $|n\rangle_k$. The extra factor $1/\sqrt{n!}$ is introduced to satisfy the normalization condition ${}_k\langle m|n\rangle_k = \delta_{m,n}$ under the bosonic commutation relation $[a_k, a_k^{\dagger}] = 1$. If one generalize the bosonic state into multi-mode states $|n_1, n_2, \ldots, n_k, \ldots\rangle$, then one can recognize that the application of the bosonic creation operator

[a] *Note*: † is complex conjugate.

yields $a_l^\dagger|\ldots,n_l,\ldots\rangle = \sqrt{n_l+1}|\ldots,n_l+1,\ldots\rangle$. Analogously, e.g. the rule can be applied to multi-mode states as

$$a_l^\dagger a_k^\dagger|\ldots,n_l,\ldots,n_k,\ldots\rangle = \sqrt{n_l+1}\sqrt{n_k+1}|\ldots,n_l+1,\ldots,n_k+1,\ldots\rangle,$$

which leads to the generalized commutation relations as

$$[a_l, a_k^\dagger] = \delta_{l,k}, \quad [a_l, a_k] = [a_l^\dagger, a_k^\dagger] = 0. \tag{3.4}$$

The states for Fermi particles also can be defined in a similar way [10, 11] with different symmetries and commutation relations.

Fourier transformation of the creation and the annihilation operators moves the particle's picture from the momentum space to the space of spatial location. The transformed operators are

$$\hat{\psi}(x) = \int_{\text{momentum}} \phi_k(x) a_k dk, \quad \hat{\psi}(x)^\dagger = \int_{\text{momentum}} \phi_k(x)^* a_k^\dagger dk, \tag{3.5}$$

where $[\{a_k, a_{k'}^\dagger\}] = \delta_{k,k'}$ depending on the particles statistics. Here $\phi_k(x)$ is arbitrary function that satisfies normalization condition $\int dx \phi_k(x)^* \phi_{k'}(x) = \delta_{k,k'}$. The transformed field operators indicate the creation and the annihilation of a particle at a position x satisfying the commutation relation in (3.2). If the particles are located in an empty box of finite volume, filled with vacuum, the particle's momentum k is quantized with discrete values. Under the circumstance, the integration is replaced by summation as,

$$\hat{\psi}(x) = \frac{1}{\sqrt{V}} \sum_k e^{ikx} a_k, \quad \hat{\psi}(x)^\dagger = \frac{1}{\sqrt{V}} \sum_k e^{-ikx} a_k^\dagger, \tag{3.6}$$

which creates and annihilates particles at the position x. The product of the creation and annihilation field operators defines the particle's density operators. The density operators

$$\hat{\rho}(x) = \hat{\psi}(x)^\dagger \hat{\psi}(x) = \frac{1}{V} \sum_{k,k'} e^{-i(k-k')x} a_k^\dagger a_{k'} \tag{3.7}$$

indicates the particle density in the volume V. Density functional theory [9] provides approaches about how the density function of a general state, not just a field in a vacuum cavity, can be obtained in the position space. The integration of the density operator can be used for the definition of the number operator

$$\int dx\hat{\rho}(x) = \sum_k a_k^\dagger a_k \equiv \hat{N}. \tag{3.8}$$

The number operator $\hat{n}_k \equiv a_k^\dagger a_k$ for the mode k is the eigenoperator of the number state $|n\rangle_k$ defined in (3.3), i.e. $a_k^\dagger a_k|n\rangle_k = n_k|n\rangle_k$. The total number of particles distributed in the different k is $\langle\hat{N}\rangle = \sum_k n_k$. In thermal equilibrium, the states have an average value of n_k which is fluctuating depend on the temperature $\beta = 1/kT$ as

$$n_k = \frac{1}{e^{\beta\varepsilon_k} - 1} \tag{3.9}$$

when the particle follows the Bose–Einstein statistics. It can be derived from the Boltzmann distribution of Bosonic particle number at a temperature T since the particle density in kth mode follows the distribution as $\rho_k = \frac{1}{\mathcal{N}}\sum_{n_k} e^{-\beta\hbar\omega_k n_k}|n_k\rangle\langle n_k|$ where $\mathcal{N}$ is normalization factor. The Bose–Einstein particle number distribution is obtained by averaging the particle number $\langle\hat{n}_k\rangle = \mathrm{Tr}[n_k\rho] = \sum_{n_k=0}^{\infty} n_k e^{-\beta\hbar\omega_k n_k}/\mathcal{N}$.

To consider more specific physical system, we discuss an example of the wave function $\psi_n(x)$ for the harmonic oscillator [12–14]. In solids, the vibrational modes of the atoms are quantized due to the canonical uncertainty principle. These quantized vibrational modes are called phonons. Phonon in solids is usually described by one-dimensional (1D) harmonic oscillators. Imagine 1D chain of N atoms joined by harmonic springs. Let x_n denote the deviation of each oscillator from its equilibrium position, ω the frequency of oscillation of each spring, and m the mass of each atoms. In the model, the lattices of atom are coupled in a harmonic way (nearest neighbor) whose Hamiltonian can be written as

$$H = \sum_n \left[\frac{p_n^2}{2m} + \frac{m\omega^2(x_n - x_{n+1})^2}{2}\right]. \tag{3.10}$$

To obtain eigenenergies, the Hamiltonian should be diagonalized to find the normal modes of the system. Once the eigenmodes are found, the system can be quantized. It allows the Hamiltonian can be written by creation and annihilation operators of the modes. To find the normal modes, Fourier transformation of the canonical variables can be taken as

$$p_n = \frac{1}{\sqrt{N}}\sum_k e^{-inka}p_k, \quad x_n = \frac{1}{\sqrt{N}}\sum_k e^{-inka}x_k. \tag{3.11}$$

In (3.10), all the other terms are trivially transformed except the interaction parts. The interaction term $\sum_n x_n x_{n+1}$ is transformed under the Fourier transform as follows:

$$\sum_n x_n x_{n+1} \to \frac{1}{N} \sum_n \sum_k \sum_l e^{inka} e^{i(n+1)la} x_k x_l \tag{3.12}$$

$$= \frac{1}{N} \sum_{kl} e^{ila} \sum_n e^{in(k+l)a} x_k x_l = \sum_k e^{ika} x_k x_{-k}.$$

Then, the Hamiltonian of the total crystal atom becomes:

$$H = \sum_k \frac{p_k p_{-k}}{2m} + \frac{m\omega^2}{2} [2(1 - \cos ka)] x_k x_{-k} \tag{3.13}$$

$$= \sum_k \frac{p_k p_{-k}}{2m} + \frac{m\Omega_k^2}{2} x_k x_{-k}, \tag{3.14}$$

where $\Omega_k^2 \equiv 2\omega^2(1 - \cos ka) = 4\omega^2 \sin^2 ka/2$. In the normal mode, the system becomes a bunch of independent harmonic oscillators. In quantum mechanics, the position and the momentum satisfy the uncertainty relation expressed by the canonical variables $[\hat{x}_k, \hat{p}_l] = i\hbar\delta_{k,l}$. Consequently, the position and momentum operators can be written by the creation and the annihilation operators as

$$a_k = \sqrt{\frac{m\omega_k}{2\hbar}} \left(x_k + \frac{i}{m\omega_k} p_{-k} \right), \quad a_k^\dagger = \sqrt{\frac{m\omega_k}{2\hbar}} \left(x_{-k} - \frac{i}{m\omega_k} p_k \right), \tag{3.15}$$

which obey the commutation relations, $[a_k, a_l^\dagger] = \delta_{k,l}$ following the uncertainty relations of the canonical variables. With the operators, the Hamiltonian can be rewritten as

$$H = \sum_k \hbar\Omega_k \left(a_k^\dagger a_k + \frac{1}{2} \right). \tag{3.16}$$

The resulting Hamiltonian describes the vibrational energy of the coupled harmonic oscillator. These collective modes of vibration are called phonons. k is the frequency of the kth mode and $a_k^\dagger$ (respectively, a_k) create (respectively, annihilate) on phonon in kth mode. Each wave vector of the state behaves independently, as a harmonic oscillator, with discrete quantum numbers, n_k, in each mode. They are the quantized version of the classical vibration in the solid, and the quanta follow the same commutation relations and Hamiltonian of the simple harmonic oscillator.

Average energy of the harmonic oscillators is discretized with a unit of Plank (black-body) constant. A black body is an idealized physical body that absorbs all incident electromagnetic radiation. Because of this perfect absorptions at all wavelengths, a black body is also the best possible emitter of thermal radiation, which it radiates incandescently in a characteristic, continuous spectrum that depends on the body's temperature. The quantized energy is emitted from a black body at a given temperature is given by

$$U = \langle H \rangle = \sum_k \hbar\Omega_k n_k = \sum_k \frac{\hbar\Omega_k}{e^{\beta\hbar\Omega_k} - 1}. \tag{3.17}$$

The average energy also can be derived when the energy follows Boltzmann distribution $p_n(\Omega_k) = e^{-n\hbar\Omega_k} / \sum_n e^{-n\hbar\Omega_k}$.

Assuming that the volume is sufficiently large, we can make the usual change from the discrete to the continuous distribution of eigenfrequencies Ω of the radiation. The number of quantum states of phonons with frequencies between Ω and $\Omega + d\Omega$ is $D(\Omega) = V\Omega^2 d\Omega/\pi^2 c^3$ [15]. In that case, the average energy in a unit volume $\langle u \rangle = U/V$ is obtained as follows:

$$\langle u \rangle = \frac{1}{V}\int_0^\infty D(\Omega)\frac{\hbar\Omega}{e^{\frac{\hbar\Omega}{kT}} - 1} d\Omega = aT^4, \tag{3.18}$$

where $a = 4\sigma/3c = \pi^2 k^4/45\hbar^3 c^3$ with Stefan–Boltzmann constant σ. Thus, the total energy of black-body radiation is proportional to the fourth power of the temperature, which is called Boltzmann's law. Subsequently, the specific heat of the radiation c_v can be obtained as

$$c_v = \frac{\partial\langle u \rangle}{\partial T} \propto T^3. \tag{3.19}$$

At low temperatures, c_v vanishes like T^3, which is consistent to the third law of thermodynamics which states that the entropy of a system at absolute zero can be taken to be zero. When the temperature is much greater than the Debye temperature, the lattice behaves classically, as indicated by statistical mechanics $c_v \simeq 3Nk$. At extremely high temperatures, the model of non-interacting phonons breaks down because the lattice eventually melt. The melting of the lattice is made possibly by the fact that the forces between the atoms in the lattice are not strictly harmonic forces. In the phonon language, the phonons are not strictly free. They must interact with each other, and this interaction becomes strong at very high temperatures.

In any case, the quantized harmonic oscillator successfully explains the low temperature behavior of a solid.

3.1.2. *Second quantization for fermion: Electron density*

If the particle is fermion, the fock space structure is need to be modified. When the identical particles are fermions, the individual occupation number n_k can only take the values 0 and 1. It is because the creation and the annihilation operators of the fermion satisfy the anticommutation relation as such,

$$\{c_k, c_l^\dagger\} = c_k c_l^\dagger + c_l^\dagger c_k = \delta_{k,l} \quad \text{and} \quad \{c_k, c_l\} = \{c_k^\dagger, c_l^\dagger\} = 0 \tag{3.20}$$

for any k and l. The second commutator can be expended to give $c_k c_l = -c_l c_k$ so that $c_k^2 = 0$ which is the only solution of $c_k^2 = -c_k^2$. It means that we can only apply the creation or the annihilation operator one time to the fock space. Physically, it is equivalent to say that any two particle cannot occupy same state at the same time which is true for the particles following the Pauli exclusion principle as like electron. It means that each mode can only have a single excitation and the fock space satisfies the relation,

$$c_k^\dagger |n_1, n_2, \ldots, n_k, \ldots\rangle = \begin{cases} (-1)^{N_k} |n_1, \ldots, n_k + 1, \ldots\rangle & \text{if } n_k = 0, \\ 0 & \text{if } n_k = 1, \end{cases} \tag{3.21}$$

where all the n_i take the value either 0 or 1, i.e. $n_i \in \{0, 1\}$. Here N_k in the exponent of the coefficient $(-1)^{N_k}$ is the particle parity number which counts the number of particles below the index k, $N_k = \sum_i^{k-1} n_i$. Applying creation operator in the other mode will generate another particle in the other mode and the sign of the wave function is determined by the number of particles in the mode between l and k as $c_l^\dagger (-1)^{N_k - N_l} c_k^\dagger$.

The density of the state for the fermion can be found in the same way for the bosonic particle using the Fourier transformation of the creation and annihilation operators. The density matrix of the Fermi particles at a position x becomes

$$\int dx \hat{\psi}(x)^\dagger \hat{\psi}(x) = \sum_k c_k^\dagger c_k = \hat{N}, \tag{3.22}$$

where the definition of the creation and the annihilation operators at the position x is identical to (3.7). At an arbitrary high temperature $\beta = 1/k_B T$, the density of the state in the fock bases is given by Boltzmann distribution of the energy values,

$$\rho(T) = \sum_k \frac{e^{-\beta c_k^\dagger c_k}}{Z}$$
$$= \sum_{n_1, n_2, \ldots} \frac{e^{-\beta(\epsilon_1 n_1 + \epsilon_2 n_2 + \cdots)}}{Z} |n_1, n_2, \ldots\rangle\langle n_1, n_2, \ldots|, \quad (3.23)$$

where n_i can only take the value either 0 or 1 and the partition function $Z = \sum_{n_1,n_2,\ldots} e^{-\beta(\epsilon_1 n_1 + \epsilon_2 n_2 + \cdots)}$ whose definition has been given in Chapter 1. Using the density of the state ρ, the expectation value of the particle number at a temperature T can be derived as $\langle \hat{n}_k \rangle = \mathrm{Tr}[\hat{n}_k \rho] \equiv n_k$

$$n_k = \frac{1}{e^{\beta \epsilon_k} + 1}, \quad (3.24)$$

which is the Fermi–Dirac distribution for the identical Fermi particle.

Electron, which is a typical Fermi particle, has spin degree of freedom in addition to the quantized momentum. It requires extra index when one represents electrons in terms of Fermi particles. Taking all of them into account, the creation and annihilation operators of electrons can be written as $c_{k,\sigma}^\dagger$ and $c_{k,\sigma}$ which satisfy the anticommutation relation in (3.20). Using the second quantized operators, one can express the density operator of free electron gas with which the effect of quantum correlation can be investigated. Consider a many fermion system with a fixed number of particles and a density matrix ρ. The elements of the reduced density matrices for $1, 2, 3, \ldots, n$ particles labeled by ρ_1, ρ_2, $\rho_3, \ldots, \rho_n$ respectively are given by [23]

$$\begin{aligned}
\langle 1|\rho_1|1'\rangle &= \langle \hat{\psi}^\dagger(1')\hat{\psi}(1)\rangle, \\
\langle 12|\rho_2|1'2'\rangle &= \langle \hat{\psi}^\dagger(2')\hat{\psi}^\dagger(1')\hat{\psi}(1)\hat{\psi}(2)\rangle, \\
\langle 123|\rho_3|1'2'3'\rangle &= \langle \hat{\psi}^\dagger(3')\hat{\psi}^\dagger(2')\hat{\psi}^\dagger(1')\hat{\psi}(1)\hat{\psi}(2)\hat{\psi}(3)\rangle, \\
\langle 1\ldots n|\rho_n|1'\ldots n'\rangle &= \langle \hat{\psi}^\dagger(n')\hat{\psi}^\dagger((n-1)')\ldots\hat{\psi}(n-1)\hat{\psi}(n)\rangle,
\end{aligned}$$

where $1 \equiv (\mathbf{r}_1, \sigma_1)$, $\mathbf{r}_1$ is the position vector and $\sigma_1 = \uparrow, \downarrow$ is the spin of the fermion. The average is given by $\langle \ldots \rangle = \mathrm{Tr}\{\rho \ldots\}$. For the sake of simplicity, all our results are illustrated at zero temperature, where $\rho = |\phi_0\rangle\langle\phi_0|$, with

$$|\phi_0\rangle = \prod_k^{k_F} c_{k,\sigma}^\dagger |vac\rangle$$

equals the ground state of the Fermi system. The $c^{\dagger}_{k,\sigma}$ is the creation operator that creates an electron of momentum k and spin σ. The Fermi momentum is denoted by k_F and the vacuum state is

$$|vac\rangle = |0, 0, 0, \ldots\rangle.$$

Moreover, $\hat{\psi}(r, \sigma) = \frac{1}{\sqrt{V}} \sum_k e^{ikr} c_{k,\sigma}$ are the annihilation operators of spin σ particle at a position r and obey the usual fermion anticommutation relations

$$\{\hat{\psi}^{\dagger}_{\sigma_1'}(\mathbf{r}_1'), \hat{\psi}_{\sigma_1}(\mathbf{r}_1)\} = \delta_{\sigma_1',\sigma_1}\delta(\mathbf{r}_1 - \mathbf{r}_1'),$$

which also take into account the spin degree of freedom σ.

As an example of electron density matrix, we illustrate a particular electronic state in a statistical mixture. With the definition of all the operators and the states of free electron, one can have a form for the density matrix for n particles which is particularly useful to investigate the properties of many electrons:

$$\rho_n = \left(1 - \sum_{ij} p_{ij}\right)\frac{\mathbf{I}}{2^n} + \sum_{ij} p_{ij}|\Psi^-_{ij}\rangle\langle\Psi^-_{ij}| \otimes \frac{\mathbf{I}}{2^{n-2}}, \tag{3.25}$$

where $|\Psi^-_{ij}\rangle = 1/\sqrt{2}(|\uparrow\downarrow\rangle - |\downarrow\uparrow\rangle)$ is the maximally entangled singlet state of the pair ij. The sum runs over all the pairs ij. The probabilities p_{ij} are functions of the relative distances between all pairs. As an example, we write down the density matrix for the two and three particle cases:

$$\begin{aligned}
\rho_2 &= p\frac{\mathbf{I}}{4} + (1-p)|\Psi^-\rangle\langle\Psi^-|, \\
\rho_3 &= (1 - p_{12} - p_{13} - p_{23})\frac{\mathbf{I}}{8} + p_{12}|\Psi^-_{12}\rangle\langle\Psi^-_{12}| \otimes \frac{\mathbf{I}}{2} \\
&\quad + p_{13}|\Psi^-_{13}\rangle\langle\Psi^-_{13}| \otimes \frac{\mathbf{I}}{2} + p_{23}|\Psi^-_{23}\rangle\langle\Psi^-_{23}| \otimes \frac{\mathbf{I}}{2},
\end{aligned} \tag{3.26}$$

where $p = (2 - 2f(r)^2)/(2 - f(r)^2)$ and $f(r) = j_1(x)/x$, with the Bessel function $j_1(x) = (\sin x - x\cos x)/x^2$ and $x = k_F r$. The relative distance between the fermion pair is denoted by r. The function $f(r)$ is one for $r = 0$ and zero for large r. For three fermions, we have three different pairs and for the pair ij: $p_{ij} = (-f_{ij}^2 + f_{ij}f_{ik}f_{jk})/(-2 + f_{ij}^2 + f_{ik}^2 + f_{jk}^2 - f_{ij}f_{ik}f_{jk})$. The function f_{ij} is a function of the relative distance between fermion i and j only. Note that the probabilities p_{ij} can be calculated for any number of particles. The state is particularly interesting because they represent the reduced state of the paired fermions in a mixture and the state can be used for the description of superconductor, which is composed of cooper pairs. Their properties will be discussed in detail at the upcoming sections.

3.2. Magnetism: Basics

When a substance is located under the influence of a magnetic field, the system suffers from a change of its internal state and behaves differently. The situation can be visualized by imagining a material which is placed near a magnetic bar. Depending upon the type of material, it is attracted (or repulsed) by the magnetic bar or sometimes remained intact. It is due to the magnetic property of the material which is mainly determined by the internal structure of the molecular states in the medium.

Magnetic field which is induced inside a material is described by two parts. One is the magnetic field intensity H from the external source and the other is the magnetization M in the material. The magnetic field intensity H describes the applied magnetic field in the location by an external magnetic source. The magnetization M denotes the induced magnetic field inside the medium due to the dipole moments which is generated by the accumulated effect of the circular motion of a charged particle. Since the magnetization M is induced by the applied field H, it is possible to assume that M is proportional to H as $\mu_0 M = \chi H$[b] where the proportionality constant χ is known as the *magnetic susceptibility* of the medium. Together with the permeability in free space $\mu_0 = 4\pi \times 10^{-7} N/A^2$, the total induced magnetic field B becomes

$$B = \mu_0 H + \mu_0 M = (\mu_0 + \chi)H = \mu H, \tag{3.27}$$

where μ is the permeability of the medium. The induced magnetic field B is proportional to the applied magnetic field whereas the strength of B varies for the different materials characterized by the permeability μ.

Depending upon the magnetic susceptibility χ, materials are grouped into a couple of different classes. Ferromagnetism is the strongest and most familiar type of magnetism which occurs at a very large χ (the order of $10^5\,\mathrm{cm}^{-3}$ and above). It is responsible for the behavior of permanent

[b]The formula adopted SI unit. In the CGS unit, the relation between the magnetization and the magnetic field becomes $\mu_0 M = \chi H$.

magnets, which produce their own persistent magnetic fields, as well as are attracted by the other magnet. If a material has reasonable positive value of χ, it is attracted by the applied magnetic field and the material is identified as *paramagnetic*. Those materials for which χ is negative are repulsed by a magnetic field and they are called as *diamagnetic*. The fact that the ions have incomplete atomic shells cause their paramagnetic behavior through the induced magnetic dipoles while the substances have atoms or ions with complete shells demonstrate the diamagnetic behavior due to the distortion in the orbital motion of the atomic systems [7]. Others have a much more complex relationship with an applied magnetic field. Substances that are negligibly affected by magnetic fields are known as non-magnetic substances.

The magnetic state (or phase) of a material also depends on temperature (and other external variables such as pressure and applied magnetic field) so that a material may exhibit more than one form of magnetism depending on its temperature, etc. It implies that the change of material's magnetic properties follows the conditions caused by the external parameters similarly to the density of a physical system which is the sensitive to pressure and temperature. In this section, we will inspect how such magnetism affects to the statistical properties of the material as the external parameters are changed. From the microscopic model, the temperature dependence of the magnetic susceptibility χ is derived and the condition of the phase transition in magnetic properties (from paramagnetism to ferromagnetism) will be discussed.

3.2.1. *Paramagnetism*

From the microscopic point of view, all the materials are composed of small segment of magnets because every atoms which constitute it behave like a little magnet. The effect is caused due to an electron in the outmost shell which is rotating around the nucleus in a circular orbit and the electronic motion generates a magnetic dipole moment perpendicular to the plane enclosed by the loop. Assuming that the electron carries the charge e moving around the orbit of area A then the magnetic moment m due to the motion is given by $m = IA$ where I is the current flowing in the closed loop. The electric current I equals to the charge passing by any point on the loop per unit time. The current is given by $I = -e\omega/2\pi$ when the electron is rotating with a constant angular velocity ω. Substituting the electric current to the magnetic moment, it becomes

$$m = -\left(\frac{e\omega}{2\pi}\right)(\pi r^2) = -\frac{e}{2m_e}L,$$

where m_e is the mass of electron and $L = m_e r^2 \omega$ is the angular momentum of the moving electron.

The result remains true in quantum mechanics except that the angular momentum L is quantized with z-directional orbital quantum number l_z. It means that, when a certain strength of magnetic field is applied to the atom, the value of the atomic angular momentum in the given direction, say z direction, is discretized and is proportional to the integer number l_z. The quantized angular moment is $L_z = \hbar l_z$ [16] and the quantum number l_z can have any integer value between $-l$ and l, i.e. $l \in \{-l, -l+1, \ldots, l-1, l\}$. Here l is another quantum number describing the total angular momentum L as $L = \hbar\sqrt{l(l+1)}$. They are the result of the solution for the Schrödinger equation having the potential energy which follows the inverse square law ($V \propto 1/r^2$). The quantization rule states that L_z is the z-component of the total angular momentum L.

The energy E of the magnetic moment which is placed in a magnetic field H depends on the orientation of the magnetic moment with respect to the applied field either in classical electrodynamics and in quantum mechanics as $E = \vec{m} \cdot \vec{H}$ where $\vec{m}$ is the magnetic moment vector and $\vec{H}$ is the applied magnetic field. In quantum mechanics, the energy is quantized due to the quantization of the angular momentum and the direction of applied magnetic field determines the z-directional angular moment. It leads to the quantized energy as

$$E = -\frac{e}{2m_e}(\hbar l_z)H = \mu_B l_z H, \tag{3.28}$$

where μ_B is defined as the Bohr magneton. $\mu_B = \frac{e\hbar}{2m_e}$ can be regarded as the quantized unit of electromagnetic moment.

In addition to the orbital angular momentum of the electrons, the spin angular momenta of the atom contribute to the total magnetic moment of the atomic state. In that case, the total magnetic moment m_T of the atom can be expressed in terms of the total angular momentum J as

$$m_T = g\left(-\frac{e}{2m_e}\right)J, \tag{3.29}$$

where g is a constant known as the *Landè g-factor*. Its value depends on the relative orientations of the orbital and spin angular momenta and the derivation of the factor can be found in the standard quantum mechanics text book [16].[c] When a magnetic field is applied to the atom, the spin degree of freedom responds to the external magnetic field and induces energy splitting called a Zeeman splitting. The Zeeman energy is a function of the quantum number of the z-directional total angular momentum as

$$E(j_z) = -\vec{m}_T \cdot \vec{H} = g\mu_B j_z H, \tag{3.30}$$

where $\vec{m}_T$ is the total magnetic moment and j_z is the z-directional quantum number for the total angular momentum. Similar to the orbital angular momentum, the z-component of the total angular momentum takes any integer value between $-j$ and j, when the total angular momentum is given by the quantum number j, i.e. $J = \hbar\sqrt{j(j+1)}$.

When the quantization condition is applied, the temperature dependence of the magnetic susceptibility $\chi(T)$ can be obtained. Assume that there are N independent paramagnetic ions per unit volume. In a finite temperature T (parameterized by $\beta = 1/kT$), the number of ions $\Delta N(j_z)$, having the magnetic moment component of m_T along the field direction j_z, follows the Boltzmann distribution as

$$\Delta N(j_z) = \mathcal{N} e^{-\beta E(j_z)}, \tag{3.31}$$

where $E(j_z)$ is the Zeeman energy in (3.30) and $\mathcal{N}$ is the normalization condition constrained by $\sum_{j_z} \Delta N(j_z) = N$. Through the evaluation of the summation, the constraint leads the condition to $\mathcal{N} = N \sinh[x/2]/\sinh[(j+1/2)x]$ where $x = g\mu_B H/kT$. Similar to that, the partition function Z which is the summation of all the Boltzmann distribution can be obtained as

$$Z = \sum_{j_z=-j}^{j} e^{-\beta E(j_z)} = \frac{\sinh[(j+1/2)x]}{\sinh[x/2]} \tag{3.32}$$

[c]For an electron spin $g = 2.0023$, usually taken as 2.00. For a free atom, the g factor is given by the Landè equation,

$$g = \frac{3}{2} + \frac{s(s+1) - l(l+1)}{2j(j+1)},$$

where s is the spin quantum number.

and x is same with x in $\mathcal{N}$. In that case, the net magnetization M in the field direction can be found by averaging the total magnetization over all the allowed energy levels which is the same as the partial derivative of free energy $F = kT \ln Z$ with respect to the magnetic field. Since the normalization factor and the partition function can be related as $Z = N/\mathcal{N}$, it is not difficult to relate the magnetization to free energy as

$$M = \sum_{j_z=-j}^{j} \mathcal{N} m_T e^{-\beta E(j_z)} = \frac{N}{Z} \sum m_T e^{-\beta E(j_z)} = \frac{\partial F}{\partial H} = N\langle m_T \rangle,$$

where m_T is the strength of the total magnetic moment $m_T = -g\mu_B j_z$ and the energy $E(j_z) = -m_T H$. Together with all the parameters of total magnetic moment and the statistical distribution, the net magnetization becomes

$$M = \sum_{j_z=-j}^{j} m_T \Delta N(j_z) = \frac{\partial F}{\partial H} = N g\mu_B j \mathcal{B}_j(x), \tag{3.33}$$

where the free energy $F = kT \ln Z$ and $\mathcal{B}_j(x)$ is referred to as the Brillouin function taking the form $\mathcal{B}_j(x) = [\frac{2j+1}{2j} \coth(\frac{2j+1}{2} x) - \frac{1}{2j} \coth \frac{x}{2}]$ where $x = g\mu_B H/kT$. The behavior of the function for large x is that $\mathcal{B}_j(x) \approx 1$ and, for small x, the Brillouin function can be approximated as $\mathcal{B}_j(x) \approx (j+1)x/3$.[d] When the total angular moment $j = 1/2$, $\mathcal{B}_{1/2}(x) = \tanh x$. This is case that there are only two energy levels in the splitting.

The asymptotic behavior of the magnetization implies that, in the low temperature limit, $x = g\mu_B H/kT \gg 1$, the net magnetization is independent to temperature T and is saturated to the value which is proportional to the quantum number j for the total angular moment as

$$M_{\text{low } T} = N g\mu_B j. \tag{3.34}$$

Contrary to that, in the high temperature limit, $x = g\mu_B H/kT \ll 1$, the net magnetization becomes inversely proportional to the temperature T as

$$M_{\text{high } T} = \frac{N(g\mu_B)^2 j(j+1)H}{3kT}. \tag{3.35}$$

---

[d] Taylor expansion of the coth function up to the third power is $\coth x = \frac{1}{x} + \frac{x}{3} - \frac{x^3}{45} + O(x^4)$. The expansion gives the asymptotic behavior of the Brillouin function for the small x.

In that region, the paramagnetic susceptibility χ is also the function of temperature,

$$\chi(T) = \mu_0 \frac{\partial M}{\partial H} = \frac{N\mu_0 (g\mu_B)^2 j(j+1)}{3kT}, \tag{3.36}$$

which is known as the Curie law. It states that the magnetic susceptibility χ is inversely proportional to temperature with the material-dependent constants. The behavior is identical to the classical results for the susceptibility and it agrees well with experimental data in the high temperature regime [17].

In fact, the relation was firstly discovered experimentally by fitting the results to a correctly guessed model by a French physicist Pierre Curie. It only holds for high temperatures, or weak magnetic fields. As the derivation has been showed, the net magnetization and the magnetic susceptibility saturate in the opposite limit of low temperatures, or strong fields case.

The external magnetic field affects to various physical properties of the materials, not only the magnetic susceptibility but also the temperature-dependent heat capacity. The heat capacity C at a fixed magnetic field can also be evaluated for the two-level system $j = 1/2$. With an energy splitting $\Delta = m_T H = g\mu_B H$ and the energy expectation value $\langle E\rangle = -\frac{1}{Z}\frac{\partial Z}{\partial \beta}$, the heat capacity becomes

$$C = \left(\frac{\partial \langle E\rangle}{\partial T}\right)_{\Delta = \text{const}} = k\frac{(\Delta\beta/2)^2}{\cosh^2(\Delta\beta/2)}.$$

The function is plotted in Fig. 3.1. In the high temperature limit, the heat capacity approaches to zero. It is because, in the high temperature, the magnet is completely randomized and is disorganized so that it cannot absorb heat. The heat capacity is peaked at a certain value of T/Δ which is called as Schottky anomaly. Since the energy splitting Δ is a linear function of external magnetic field, the role of the magnetic field is identical to temperature in the change of the heat capacity.

At the opposite limit, $T \to 0$, the state becomes pure ground state and all the magnetic moments are order in a way that heat cannot be accommodate in the system neither. It is consistent with the third law of thermodynamics state that the entropy S becomes constant at the absolute zero temperature. Following the third law,

$$S \to 0, \quad C(T) = \frac{dQ}{dT} = T\frac{dS}{dT} \to 0,$$

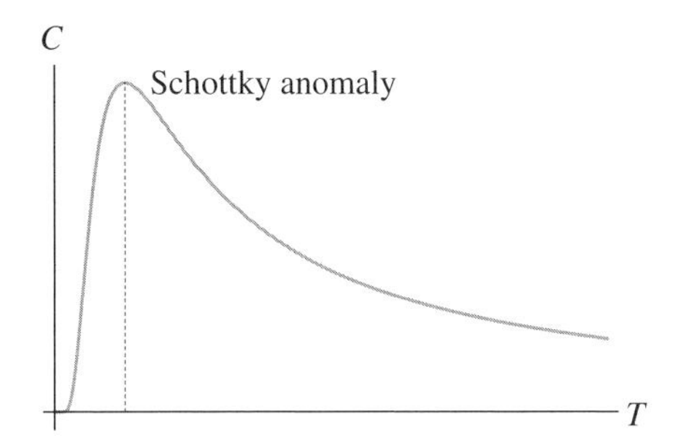

Fig. 3.1. Heat capacity of a two-level system as a function of T, where Δ is the energy splitting. The heat capacity disappears in the limits, $T \ll 1$ and $T \gg 1$. It is notable that the external magnetic field $H \propto \Delta$ contributes to the change of heat capacity in the same way of inverse temperature.

thus the heat capacity goes zero at an arbitrary low temperature. Same effect can be observed when the applied external field is extremely strong. However, the low temperature behavior of the magnetic material does not well agreed with the theory of magnetism (paramagnetism). It is where the quantum theory of magnetic dipole interaction becomes important. We discuss the effect in the section on the ferromagnetism.

3.2.2. *Ferromagnetism: Ising model*

Ferromagnetism refers to solids that are magnetized without an applied magnetic field. These solids are said to be spontaneously magnetized. Ferromagnetism occurs when paramagnetic ions in a solid lock together in such a way that their magnetic moments are all aligned (on the average) in the same direction. At high enough temperatures, this locking breaks down and ferromagnetic materials become paramagnetic.

The effect is originated from the interaction between the spin degree of freedom of the ions. Consider a magnetic moment with a concentration of ions containing spins. Given an internal interaction tending to line up the magnetic moments parallel to each other. The interaction is called as the exchange field[e] which arises due to the electrostatic interaction between the electrons. The orienting effect of the exchange field is opposed by thermal excitation, and the spin order is destroyed at a high temperature.

[e]Also called the molecular field or the Weiss field, after Pierre Weiss who was the first to imagine such a field. The exchange field simulates a real magnetic field in the expressions for the energy.

The magnetization is defined as the magnetic moment per unit volume and the value is the function of the temperature T in thermal equilibrium as is discussed. If domains (regions magnetized in different directions) are present, the magnetization refers to the value within a domain. In the mean-field approximation (the detailed theory will be discussed later), we assume each magnetic atom experiences a field $H_{\rm av}$ proportional to the average magnetization:

$$H_{\rm av} = \lambda M, \tag{3.37}$$

where λ is a constant, independent of temperature. The relation states that each spin sees the average magnetization of all the other spins which is the concept of the mean-field approximation. In fact, as it is true in the most of the case, the spins may see only nearest neighbors, but our simplification is good for a first look at the problem.

Together with the applied external magnetic field $H_{\rm app}$, the average magnetization generates extra magnetic field $H_{\rm av}$ at the every sites of ion. Then the spin experiences the total magnetic field $H_{\rm app} + H_{\rm av}$ such that the magnetization of the spin becomes

$$\mu_0 M = \chi(H_{\rm app} + \lambda M), \tag{3.38}$$

where χ is the paramagnetic susceptibility. In the case of ferromagnetic case, the susceptibility is defined from (3.38) since

$$\mu_0 M = \frac{\mu_0 \chi}{\mu_0 - \lambda\chi} H_{\rm app} \equiv \chi_f H_{\rm app}, \tag{3.39}$$

and the ferromagnetic susceptibility is related with the paramagnetic susceptibility as $\chi_f = \frac{\mu_0\chi}{\mu_0 - \lambda\chi}$. The temperature dependence of the paramagnetic susceptibility has been derived in (3.36) and it lead to the temperature dependence of ferromagnetic susceptibility as

$$\chi_f(T) = \frac{c}{T - T_0} \tag{3.40}$$

with, so-called, the *Curie constants*, $c = N\mu_0(g\mu_B)^2 j(j+1)/3k_B$ and the *Curie temperature*, $T_0 = \lambda c/\mu_0$. It implies that the ferromagnetic susceptibility has a singularity at T_0. At this temperature T_0 (and below), there exists a spontaneous magnetization, because if χ_f is infinite we can have a finite M for zero H_{app}. The behavior (3.40) is known as the Curie–Weiss law and describes fairly well the observed susceptibility variation in the paramagnetic region above the Curie point.

If the temperature is above T_0, the spontaneous magnetization vanishes. It separates the disordered paramagnetic phase at $T > T_0$, from the ordered ferromagnetic phase at $T < T_0$. In that sense it can be said that the paramagnets are disordered phase. This is because atoms do not magnetically interact with each other and as soon as the little magnets in a solid are coupled to each other we can have an establishment of order. It gives rise to the phase transition:

Disorder → Order (phase transition)

and the phase transition occurs at the Curie temperature T_0. The critical temperature can also be related with the interaction of the spins in the microscopic model. It is the Ising model which is a simplified microscopic description of ferromagnetism. The mean field approximation of the Ising model can relate the interaction strength and the critical temperature for the phase transition.

Ising model is based on the spin–spin interactions. The charge distribution in a system of two electrons depends on whether the electron spins are parallel or antiparallel. The Pauli principle excludes two electrons of the same spin from being at the same place simultaneously but not the two electrons of opposite spin. It also implies that the distance between two parallel spins is larger than the antiparallel spins. Thus, as a result of the inverse distance law of the energy, the antiparallel spins has larger energy than the parallel spins. In that situation, the electrostatic energy of a system will depend on the relative orientation of the spins and the difference in energy defines the exchange energy.

Consider the spin directions valued as $S = \pm 1$. It means that any spin either can point out one direction or the opposite. The exchange field gives an approximated representation of energy in the spin–spin interaction. On the assumptions, it can be shown that the energy of interaction of spins S_i and S_j becomes

$$E_{\text{int}} = -J \sum_{i \neq j} S_i S_j, \tag{3.41}$$

where J is the exchange integral and is related to the overlap of the charge distributions of the atoms at i and j. The model is called the Ising spin model and the quantum mechanical generalization of the model is known as the Heisenberg model.[f] We will explain about the Heisenberg model more in detail in the later section. The exchange energy of two electrons is written in the form (3.41), just as if there were a direct coupling between the directions of the two spins. For many purposes in ferromagnetism, it is a good approximation to treat the spins as classical angular momentum vectors.

If one use the mean field theory, it is not difficult to identify the statistical distribution of the spin orientation. Instead of considering the exact spin–spin interactions, it is possible to approximate the interaction energy to the average spin magnetization. It is because the actual field that each spin experiences can be approximated to the average magnetic field at the position of the spin. With the mean field assumption, the exchange energy is approximated as

$$E_{\text{int}} \approx E_{\text{MF}} = JS \sum_{i=1}^{N} S_i, \tag{3.42}$$

where S is the average magnetic moment of the spins, i.e. $S = \langle S_j \rangle = \text{Tr}[S_j e^{-\beta E_{\text{MF}}}]/Z$ with the partition function $Z = \text{Tr}\, e^{-\beta E_{\text{MF}}}$.[g] In fact, the spin magnetic moment corresponds to the average magnetic field $H_{\text{av}} = \langle S_j \rangle$ which is the linear function of magnetization M. The definitions of mean field energy (3.42) and the magnetization (3.37) result in a self-consistent equation:

$$S = \frac{\sum_{S_1,S_2,\ldots=-1}^{1} S_j e^{-\beta E_{\text{MF}}}}{\sum_{S_1,S_2,\ldots=-1}^{1} e^{-\beta E_{\text{MF}}}} = -\tanh\left(\frac{JS}{k_B T}\right). \tag{3.43}$$

It can be seen that the equation above has non-trivial solution only when the coefficient of S in the tangent hyperbolic function is larger than one. The behavior is sketched in Fig. 3.2. The direct result is that the ferromagnetism occurs when

$$\frac{J}{k_B T} \geq 1 \;\Rightarrow\; T_c = \frac{J}{k_B}.$$

From the equality, the critical temperature can be derived from the interaction strength between the spins. This is quite a general result:

[f] In the Heisenberg model, the spin values Si and S_j are replaced by Pauli spin operators.
[g] Here Tr means summing over all the S_j values, which are $+1$ and -1. Therefore, the spin S_j can be defined by 2×2 diagonal matrix, as $S_j = \text{diag}[1, -1]$ which has same matrix representation with Pauli spin-z matrix.

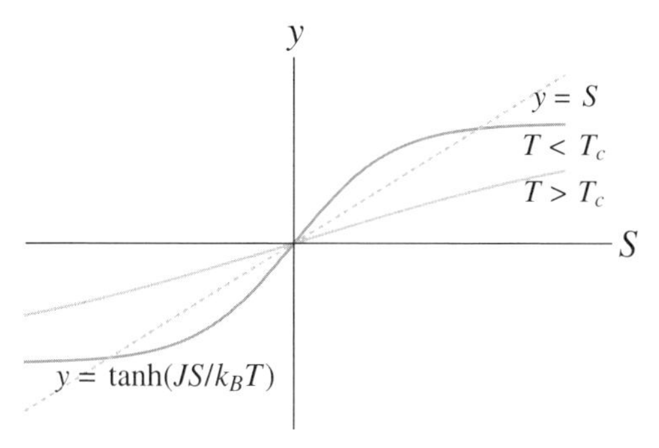

Fig. 3.2. Graph for the solution of self-consistent equation. The linear function $y = S$ and the tangent hyperbolic function $y = \tanh(JS/kT)$ are plotted together. The tangent hyperbolic function will cross over the linear function only when the coefficient of S, J/kT, is larger than one.

Critical behavior usually occurs when T_c is smaller than typical interaction strength. However, the ferromagnetic phase transition in the Ising model is possible in a specific geometric configuration of the spin. The condition for the phase transition will be discussed in Section 3.2.3.

3.2.3. *Peierls argument: Phase transition*

A phase transition [18] is a macroscopic effect which occurs when a parameter of the system is tuned across a critical value, beyond which the system takes on a qualitatively different phase. For instance, an ordered phase could be a magnetized piece of metal or Bose–Einstein condensate (BEC) in a Bose-gas. The notion of phase transition is intimately related to that of criticality which is a concept of central importance in solid-state physics. Macroscopic objects, in the form of solids, liquids and gases, undergo a diverse range of phase transitions under variation of the temperature, or external field, or pressure for instance.

When the 2D Ising spin lattice undergoes a phase transition at its critical temperature, this means that above this temperature, the spins were in a randomized, disordered state, where all directions are equally likely, but below that temperature all the spins align and point in some (randomly chosen) direction, see upper schematic in Fig. 3.3. This transition from a disordered to an ordered state is signified by spontaneous symmetry breaking of the z-directional invariance of the state of the Ising lattice. An additional parameter has to be introduced to specify which particular "order" was chosen by the system. The *order parameter* in this case is

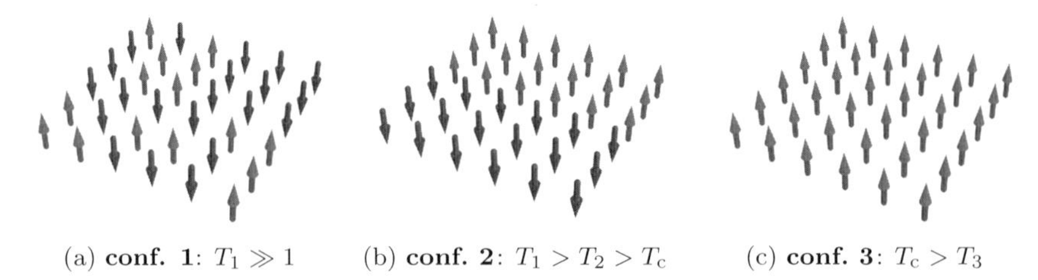

(a) **conf. 1**: $T_1 \gg 1$ (b) **conf. 2**: $T_1 > T_2 > T_c$ (c) **conf. 3**: $T_c > T_3$

Fig. 3.3. Graphical configuration of spins in the 2D Ising model. It shows the formation of an ordered (magnetic) phase of spins with decreasing temperature in the 2D Ising model. **conf. 1**: $T_1 \sim$ big: Spins are randomly flipped so that half of them point up and half of them down, without any kind of order. The magnetization is zero, $M_1 = 0$. **conf. 2:** $T_2 > T_c$: Spins in the lower half and upper half point in opposite directions and form a Weiss-domain each with distinct magnetization. Nevertheless, the total magnetization is still zero. **conf. 3:** $T_3 \to 0$: All spins point in one direction and form one ordered phase with $M_3 = M_{\max}$. The critical temperature, T_c, for this phase transition is passed somewhere in between **conf. 2** and **conf. 3**.

the total magnetization, which is finite below the point of criticality and becomes suddenly zero at and above this point. Symmetry breaking is, therefore, another indicator of phase transitions.

The main advantage of the concept of criticality is the fact that wide ranging systems behave very similarly to each other at the critical point. This realization leads to the concept of universality, namely that correlations, and consequently other observable system properties can, when critical, be described using only a small number of parameters, that are completely independent of the nature of the system. As a result, one can derive some rather general conclusions about the existence of phase transitions. It is well known that there are no discrete or continuous order parameter phase transitions in 1D, no matter what particular system we are talking about, providing that the interactions are short ranged. This result has been proven using many different methods, first by Peierls (and tidied-up by Griffiths) [24], then by Mermin and Wagner [25], and finally by Hohenberg [26]. The gist of all these arguments is that any thermal fluctuations in 1D are enough to destroy order that is generated by short-range interactions. In other words, entropy always dominates energy in 1D. In this section, we will illustrate the Peierls argument about the phase transition.

The intuitive picture of Peierls' argument [24] is as follows. The classical 1D spin chain interacting with the Ising short-range coupling would undergo a phase transition from a disordered phase to an ordered phase if a non-zero magnetization arose below a certain critical temperature. To test whether the ordered state can be an equilibrium state of the system for any finite temperature, we assume that the system is initially ordered and then perturb the state slightly while observing the change of the free energy, ΔF, with

$$F = U - T\,S, \tag{3.44}$$

where U is the internal energy, T the temperature and S the entropy of the state. If the ordered state was an equilibrium state, then its free energy must be minimal and any perturbation out of this state would increase free energy, $\Delta F > 0$. If, on the other hand, there exists a perturbed configuration that can decrease the free energy, the system will tend towards this new configuration preferring it energetically over the ordered state. This immediately implies that any order will be destroyed and no criticality is exhibited.

The Hamiltonian of the 1D Ising spin chain with nearest neighbor interaction is given by

$$H = -J \sum_{j=1}^{N} \sigma_j \sigma_{j+1}, \tag{3.45}$$

where the σ_j are the Pauli spin operators in z direction for the jth qubit and N is the number of qubits in the chain. The matrix representation of the Pauli operator will be discussed in the next section. The interaction J is positive in the ferromagnetic case and negative in the antiferromagnetic case. The ordered state is reached when all spins are aligned, either in the positive or negative z direction. We proceed by applying a single random flip, which realizes the smallest possible perturbation. The flipping spin is sketched in Fig. 3.4. The change of internal energy under such perturbation is $\Delta U = \langle H \rangle_{\text{flipped}} - \langle H \rangle_{\text{ordered}} = -J(N-4) - (-JN) = 4J$, because every spin is connected to two nearest neighbors. There are N different ways to flip a single spin so that the change in entropy is $\Delta S = k_B \ln N$. Therefore the change in free energy is

$$\Delta F = 4J - k_B T \ln N, \tag{3.46}$$

and since we are looking at the thermodynamical limit of large N it is clear that ΔF will always be negative for any finite temperature. The conclusion is that in 1D no ordered state can exist in a finite temperature equilibrium. The entropy, i.e. the thermal fluctuations, will always outweigh the tendency

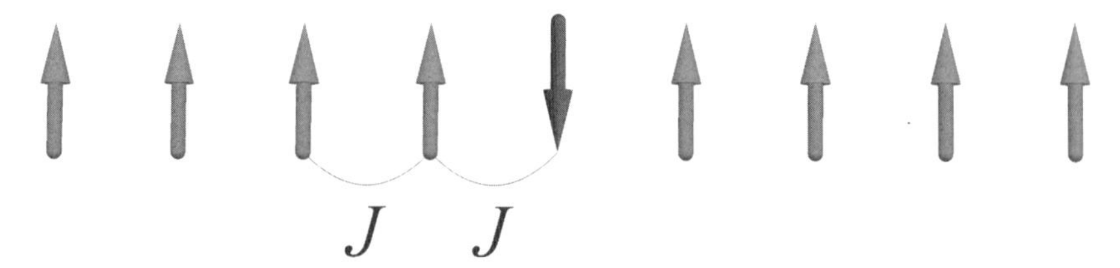

Fig. 3.4. Schematic diagram of the spin interaction in 1D spin chain. It is assumed that they are all lined up in a same direction initially. By flipping one of the spins, the free energy is changed scaled with the total number of spins.

of energy to create order. Similar conclusions can be drawn for all short-range interactions completing Peierls' proof that no phase transition takes place in 1D.

This situation is changed dramatically in two and higher dimensions. In 2D the ordered state is also the one where all the spins point in the same direction. However, flipping a single spin is now no longer enough to destroy this order; instead the smallest perturbation must cut the lattice into two different domains.[h] Let us now calculate the best case scenario for such a cut to destroy order. As far as the balance between change in energy and entropy is concerned, it is the one which cuts the lattice into two halves by crossing as many nearest neighbor couplings as possible along the way. For such a cut, the change in energy is at most $2NJ$, since this is the number of couplings present in the lattice (see Fig. 3.5 for an illustration). The entropy change, on the other hand, is roughly $k_B \ln 3^N$ because after each spin we have a choice to proceed with the cut in at most three possible directions (everywhere but backwards). The change in free energy for this kind of perturbation is thus

$$\Delta F = 2NJ - k_B T N \ln 3. \tag{3.47}$$

By setting $\Delta F = 0$ we obtain the following critical temperature for the stability of the ordered phase

$$T_{\text{crit}} = \frac{2J}{k_B \ln 3}. \tag{3.48}$$

This is astonishingly close to the exact temperature obtained by Onsager [90] ($T = 2.27\frac{J}{k_B}$), especially, given that the above argument is very simplistic.

[h] If we were to flip only single spins here and there the 2D lattice would still be connected as one phase and in the thermodynamic limit these errors would not matter. To properly bound or separate a 2D object, one needs a 1D line.

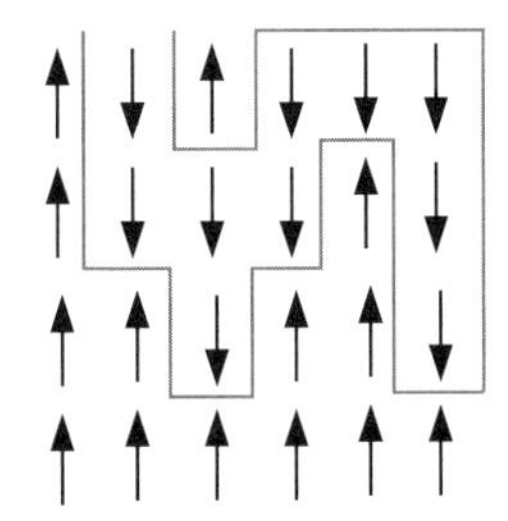

Fig. 3.5. How to flip spins in a 2D lattice in the best case scenario. This figure shows a possible flipping pattern that cuts the lattice into domains of equal alignment. A path that wiggles back and forth inverts a lot of nearest neighbor couplings and the energy of the lattice increases considerably (maximally). However there are many similar, albeit different such paths so that the entropy of the lattice increases at the same rate as the energy. Therefore, there exists a finite temperature range, where the entropic fluctuations are more costly than the energy gain and the fully aligned (ordered) configuration remains the minimum of free energy. Above this critical temperature, the ordered phase is destroyed and a phase transition occurs.

To summarize, in 2D the two competing quantities can be balanced and a phase transition can occur at a finite temperature. With the same argument one can show that this holds also true for higher dimensions.

3.3. The Hubbard Model

In Section 3.2, we have discussed the theory of magnetism in the high temperature regime where the system is intrinsically classical. When a magnetic material is in an arbitrary low temperature, the electrons in the material subjected to the exchanged interactions follow the law of quantum mechanics.

The Hubbard model is an approximate model used to describe the insulator–conductor transition systems, ferromagnetic material and superconducting effects. The Hubbard model, named after British physicist John Hubbard, is the simplest model of interacting particles in a lattice, having two parts in the Hamiltonian: a kinetic term allowing for tunneling ("hopping") of particles between sites of the lattice and a potential term consisting of an on-site interaction. The particles can either be fermions, as in Hubbard's original work, or bosons, when the model is referred to as either the "Bose–Hubbard model" or the "boson Hubbard model".

The model was originally proposed to describe the energy of the itinerant electrons in solids and has been the focus of particular interest as a model for high-temperature superconductivity. The theory of superconductivity will be given in the subsequent section. More recently, the Bose–Hubbard model has been used to describe the behavior of ultracold atoms trapped in optical lattices [95]. Recent ultracold atom experiments [93, 94] have also realized the original, fermionic Hubbard model in the hope that such experiments could yield its phase diagram. The model is extensively studied and there is an review article by Lieb [91] which covers wider topics on the model.

3.3.1. *Description of the model*

Consider a crystal of lattice sites with a total of N itinerant electrons hopping between the neighboring lattice sites. In that case, each site is capable of accommodating two electrons of opposite spin, with an interaction energy $U > 0$, which mimics a screened Coulomb repulsion among electrons. The Hubbard model [92] is described by the Hamiltonian

$$H = -\sum_{\langle ij \rangle} \sum_{\sigma \in \{\uparrow,\downarrow\}} t_{ij} a_{i\sigma}^{\dagger} a_{j\sigma} + U \sum a_{i\uparrow}^{\dagger} a_{i\uparrow} a_{i\downarrow}^{\dagger} a_{i\downarrow}, \tag{3.49}$$

where the first term describes electron hopping and the second term is on-site repulsion term. The subscripts i and j denote the site index and $\sigma \in \{\uparrow, \downarrow\}$ identifies the particles spin degree of freedom. $a^{\dagger}$ and a operators could be either fermionic (the usual) or bosonic (Bose–Hubbard model).

The solution of the Hubbard model has been studied by the full range of analytic technique developed by condensed matter physicist. Here, we will only attempt to explain the basic characters of the Hamiltonian and discuss the simple approach in which it is solved. It is possible to get the first insight of the Hubbard model by considering the limiting cases.

Suppose that the number of electrons is equal to the number of ionic sites. In that case, the repulsion between the electrons is very strong and the screened Coulomb interaction results $U \gg t_{ij}$. To make the problem simpler, we start from the limiting case $t_{ij} \to 0$ which implies that the electron hopping is not allowed. The state of the system is an insulator because the electron cannot move around the sites. The ground eigenstate of the system in the limit is given as

$$\prod_n a_{n\sigma}^{\dagger} |0\rangle, \tag{3.50}$$

where $|0\rangle$ is vacuum state and $a_{n\sigma}^{\dagger}$ is the creation operator at the nth site with spin σ. Here σ can be either $\uparrow$ or $\downarrow$. It means that all the sites are occupied by single electron with arbitrary spins. Due to the choices of the spin, the ground state is hugely degenerated. To have better illustration, we consider the case of two sites and two electrons. In the simple case, any of the states

$$|\psi_1\rangle \equiv a^\dagger_{1\uparrow}a^\dagger_{2\uparrow}|0\rangle, \quad |\psi_2\rangle \equiv a^\dagger_{1\downarrow}a^\dagger_{2\uparrow}|0\rangle, \tag{3.51}$$

$$|\psi_3\rangle \equiv a^\dagger_{1\uparrow}a^\dagger_{2\downarrow}|0\rangle, \quad |\psi_4\rangle \equiv a^\dagger_{1\downarrow}a^\dagger_{2\downarrow}|0\rangle \tag{3.52}$$

can be the ground state of the Hamiltonian. They are the states having equally distributed electrons in the sites 1 and 2 while all the combination of the spins are possible. Thus, the ground states have four-fold degeneracies. The eigenenergy of the states are all $E = 0$. Additionally, it is straightforward to see that the first excited states of the simple model are

$$|\psi_5\rangle \equiv a^\dagger_{1\uparrow}a^\dagger_{1\downarrow}|0\rangle, \quad |\psi_6\rangle \equiv a^\dagger_{2\uparrow}a^\dagger_{2\downarrow}|0\rangle, \tag{3.53}$$

which have the eigenenergy $E = U$. The energy gap between the ground state and the first excited state is U in the case. In fact, (3.51) and (3.53) are the full spectrum of the Hamiltonian.

Now, we turn on t_{ij} as a perturbation which is still small value compared to U. It is our question how the perturbation does change the state of the system. The hopping term allows the electrons to move from site 1 to site 2 or vice versa such that the electric current can be generated. With the hopping terms, the electrons in the sites are strongly coupled and the state becomes a conductor.

According to the perturbation theory, when there exists a small perturbation in a Hamiltonian as $H = H_0 + tV$ and $t \ll 1$, the asymptotic eigenenergies and the eigenstates are obtained from the solution of the Schrödinger equation and they are

$$E_i = E_i^{(0)} + tE_i^{(1)} + t^2E_i^{(2)} + \cdots, \tag{3.54}$$

$$|\psi_i\rangle = |\psi_i^{(0)}\rangle + t|\psi_i^{(1)}\rangle + t^2|\psi_i^{(2)}\rangle + \cdots, \tag{3.55}$$

where i is the index for the orthogonal eigenstates and the zeroth-order eigenstate is defined from the eigenequation without the perturbation, $H_0|\psi_i^{(0)}\rangle = E_i^{(0)}|\psi_i^{(0)}\rangle$. From the original eigenvalue equation, the higher-order perturbation energies and the eigenvector can be derived, for example,

$$E_i^{(1)} = \langle \psi_i^{(0)} | V | \psi_i^{(0)} \rangle, \quad E_i^{(2)} = \sum_{j \neq i} \frac{|\langle \psi_j^{(0)} | V | \psi_i^{(0)} \rangle|^2}{E_i^{(0)} - E_j^{(0)}}, \tag{3.56}$$

where V is the perturbed part of the Hamiltonian.

Using the perturbation theory, it is possible to identify the energy spectrum shift of our model when t has a small, but finite, value. In the model, the first-order hopping processes is not possible. The first-order hopping probabilities of all the eigenstates in (3.51) and (3.53) are proportional to the expectation value of the perturbed Hamiltonian and it can be evaluated with the substitution $V = \sum_{\sigma \in \{\uparrow,\downarrow\}} (a_{1\sigma}^\dagger a_{2\sigma} + a_{2\sigma}^\dagger a_{1\sigma})$, as

$$E_i^{(1)} = \langle \psi_i^{(0)} | (a_{1\sigma}^\dagger a_{2\sigma} + a_{2\sigma}^\dagger a_{1\sigma}) | \psi_i^{(0)} \rangle = 0, \tag{3.57}$$

where the subscript i of the first-order energy $E_i^{(1)}$ is originated from the corresponding choices of eigenstate $|\psi_i^{(0)}\rangle$. The zero hopping probability (3.57) means that the perturbation do not contribute to the first-order energy correction.

In the same way, the second-order correction terms can also be evaluated and the values are varied for the different eigenstates. Defining the symmetrized and antisymmetrized states in terms of the particles spins, we have two states,

$$|\psi^+\rangle = \frac{1}{\sqrt{2}}(|\psi_3\rangle + |\psi_4\rangle), \quad |\psi^-\rangle = \frac{1}{\sqrt{2}}(|\psi_3\rangle - |\psi_4\rangle), \tag{3.58}$$

which are also the orthonormal ground states of the on-site Hamiltonian. To find the second-order perturbation term, we apply the hopping Hamiltonian to the ground states (3.51), which gives

$$V|\psi_1\rangle = 0, \quad V|\psi_2\rangle = 0, \tag{3.59}$$

$$V|\psi_3\rangle = -|\psi_5\rangle - |\psi_6\rangle, \quad V|\psi_4\rangle = |\psi_5\rangle + |\psi_6\rangle. \tag{3.60}$$

It naturally identifies that the symmetric state $|\psi^+\rangle$ is the eigenstate of the hopping term because $V|\psi^+\rangle = 0$ and the antisymmetrized state $|\psi^-\rangle$ is not the eigenstate of V. After a little bit of algebra using Eq. (3.56), the second-order terms of the ground states are

$$E_1^{(2)} = E_2^{(2)} = E_+^{(2)} = 0, \quad E_-^{(2)} = -2/U, \tag{3.61}$$

where $E_1^{(2)}$ and $E_2^{(2)}$ are the second-order perturbation terms for $|\psi_1\rangle$ and $|\psi_2\rangle$ while $E_+^{(2)}$ and $E_-^{(2)}$ are for $|\psi_+\rangle$ and $|\psi_-\rangle$. It means that the state

$|\psi^+\rangle \equiv \frac{1}{\sqrt{2}}(|\uparrow_1\downarrow_2\rangle + |\downarrow_1\uparrow_2\rangle)$ has zero contribution to the second-order perturbation energy. However, the second-order contribution of the singlet $|\psi^-\rangle \equiv \frac{1}{\sqrt{2}}(|\uparrow_1\downarrow_2\rangle - |\downarrow_1\uparrow_2\rangle)$ is nonetheless trivial and the total energy of singlet becomes $E_- = t^2E_-^{(2)} = -2t^2/U$ which is ground state energy. Therefore, the two electron ground state of the two-site Hubbard model is the antisymmetrized singlet state. This mechanism is effectively equivalent to the phenomena called *superexchange*. As was briefly discussed in the ferromagnetism section, the superexchange describes the electron–electron interaction, not by the Coulomb interaction but by the identicality of two electrons. The interaction is the results of the anticommutation relation of fermions which is identical under the particle exchanges. This is purely quantum mechanical effects which do not have classical correspondence.

3.3.2. *Correspondence to Heisenberg model*

In the weak hopping limit, the Hubbard model can be approximated to Heisenberg Hamiltonian. Taking into account the spin degree of freedom only, Heisenberg Hamiltonian represents the electron interaction with the spin degree of freedom. It is similar to the Ising interaction and the resulting Hamiltonian for spin is

$$H_{\text{Heisenberg}} = -\frac{t^2}{2U}(\sigma_1^x\sigma_2^x + \sigma_1^y\sigma_2^y + \sigma_1^z\sigma_2^z - 1), \tag{3.62}$$

where σ^x, σ^y, σ^z are Pauli spin operators. The subscript of the operators is the site index. The matrix representation of the spin operators are

$$\sigma^x = \begin{pmatrix} 0 & 1 \\ 1 & 0 \end{pmatrix}, \quad \sigma^y = \begin{pmatrix} 0 & -i \\ i & 0 \end{pmatrix}, \quad \sigma^z = \begin{pmatrix} 1 & 0 \\ 0 & -1 \end{pmatrix} \tag{3.63}$$

and they represent the spin direction in Hilbert space.[i]

The meaning of $\sigma_1^x\sigma_2^x$ is the interaction between the spins in the sites 1 and 2. The operator multiplication between the different sites should

[i]Hilbert space is complex vector space with unit inner product. The eigenstates of the Pauli spin matrix compose the complete set of orthogonal bases. The bases of the matrix are $|\uparrow\rangle, |\downarrow\rangle$ and the eigenvectors for the matrices are $|+\rangle, |-\rangle$ for σ_x, $|+i\rangle, |-i\rangle$ for σ^y and $|\uparrow\rangle, |\downarrow\rangle$ for σ^z. The relations between them are $|\pm\rangle = (|\uparrow\rangle \pm |\downarrow\rangle)/\sqrt{2}$ and $|\pm i\rangle = (|\uparrow\rangle \pm i|\downarrow\rangle)/\sqrt{2}$.

be given in the tensor product which means in the matrix form

$$\sigma_1^x \otimes \sigma_2^x = \begin{pmatrix} 0 & 1 \\ 1 & 0 \end{pmatrix} \otimes \begin{pmatrix} 0 & 1 \\ 1 & 0 \end{pmatrix} = \begin{pmatrix} 0 & 0 & 0 & 1 \\ 0 & 0 & 1 & 0 \\ 0 & 1 & 0 & 0 \\ 1 & 0 & 0 & 0 \end{pmatrix}, \tag{3.64}$$

$$\sigma_1^y \otimes \sigma_2^y = \begin{pmatrix} 0 & -i \\ i & 0 \end{pmatrix} \otimes \begin{pmatrix} 0 & -i \\ i & 0 \end{pmatrix} = \begin{pmatrix} 0 & 0 & 0 & -1 \\ 0 & 0 & 1 & 0 \\ 0 & 1 & 0 & 0 \\ -1 & 0 & 0 & 0 \end{pmatrix}, \tag{3.65}$$

$$\sigma_1^z \otimes \sigma_2^z = \begin{pmatrix} 1 & 0 \\ 0 & -1 \end{pmatrix} \otimes \begin{pmatrix} 1 & 0 \\ 0 & -1 \end{pmatrix} = \begin{pmatrix} 1 & 0 & 0 & 0 \\ 0 & -1 & 0 & 0 \\ 0 & 0 & -1 & 0 \\ 0 & 0 & 0 & 1 \end{pmatrix}. \tag{3.66}$$

Using the matrices, the total matrix of the Heisenberg Hamiltonian can be found as

$$H_{\text{Heisenberg}} = \frac{t^2}{2U} \begin{pmatrix} 0 & 0 & 0 & 0 \\ 0 & -2 & 2 & 0 \\ 0 & 2 & -2 & 0 \\ 0 & 0 & 0 & 0 \end{pmatrix} \tag{3.67}$$

and the eigenvalues of the total matrix are $0, 0, 0, -2t^2/U$. The corresponding eigenstates for the first excited states are

$$|\uparrow\uparrow\rangle, \quad |\downarrow\downarrow\rangle, \quad \frac{1}{\sqrt{2}}(|\uparrow\downarrow\rangle + |\uparrow\downarrow\rangle) \tag{3.68}$$

and the non-degenerated ground state is

$$\frac{1}{\sqrt{2}}(|\uparrow\downarrow\rangle - |\uparrow\downarrow\rangle). \tag{3.69}$$

The energy spectrum of the Hamiltonian is sketched in Fig. 3.6. The singlet state is the unique ground state and the first excited state is degenerated with three states which are the triplet states.

They are the same energy spectrums with the two electrons confined in the two ions and the energy difference between the ground state and the first excited state is $\Delta E = -\frac{2t^2}{U}$. The ground state of two-site Heisenberg

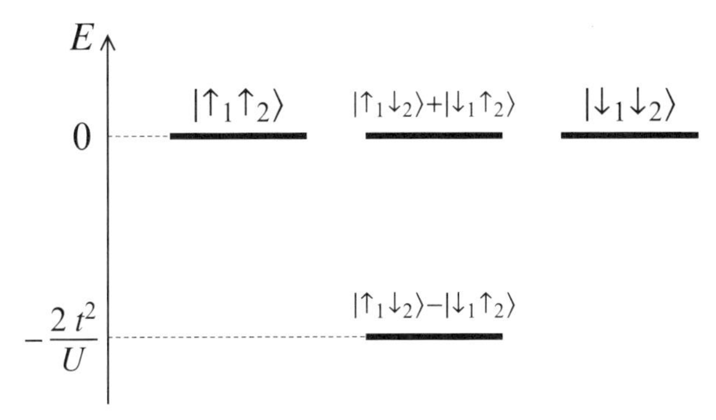

Fig. 3.6. Energy spectrum of the Heisenberg Hamiltonian for the two spins. Ground state is singlet state while the excited states are three triplet states. The excited state is three-fold degenerated and the energy gap is $\Delta E = 2t^2/U$.

model is singlet state in which the electrons are superposition state with antiferromagnetic order. The state is conducting phase but the state does not show magnetic properties.

When the state of two electrons is written as (3.68) and (3.69), one should be careful about the fact that the electrons are confined in the potential of two ions. The triplet states (3.68) are symmetric for the particle exchange while the singlet state (3.69) is antisymmetric. It is because the notation take into account the spin degree of freedoms only. Based upon the Heitler–London's theory of two Hydrogen molecules [119], the complete wave function of the electrons is given together with the parameters for the spatial coordinate as it satisfies the electron's anticommutation relations. With the consideration, the singlet wave function is

$$|\psi^-(r_1,r_2)\rangle = \frac{1}{\sqrt{2}}[\phi_1(r_1)\phi_2(r_2) + \phi_1(r_2)\phi_2(r_1)]\frac{1}{\sqrt{2}}(|\uparrow_1\downarrow_2\rangle - |\uparrow_1\downarrow_2\rangle), \tag{3.70}$$

whereas the wave function of the triplets are

$$|\psi^+(r_1,r_2)\rangle = \frac{1}{\sqrt{2}}[\phi_1(r_1)\phi_2(r_2) - \phi_1(r_2)\phi_2(r_1)]\begin{cases} |\uparrow_1\uparrow_2\rangle, \\ \frac{1}{\sqrt{2}}(|\uparrow_1\downarrow_2\rangle + |\uparrow_1\downarrow_2\rangle), \\ |\downarrow_1\downarrow_2\rangle, \end{cases} \tag{3.71}$$

where $\phi_i(r_i)$ is the spatial part of the electron's wave function confined in the nucleus with Coulomb potential. However, it is enough to consider the spin parts only as far as the magnetic properties are concerned. The discussion gives us a qualitative understanding of the Heisenberg Hamiltonian and we will discuss about the Heisenberg model more in detail in the later section.

3.4. The Kondo Effect: Magnetic Impurity

The electrical resistance of a pure metal usually drops as its temperature T is lowered. It is because electrons can travel through a metallic crystal more easily when the vibrations of the atoms are small. However, there are examples of the resistance saturation as the temperature is lowered below about 10 K. It has been known that the behavior is caused by static defects in the material. In addition, there is experimental observation of an increase of resistivity below certain temperature T. The effect, explained by Japanese physicist, Kondo [99] in 1964, is due to magnetic impurity in the material interacting with neighboring electron spins.

Kondos theory correctly describes the observed upturn of the resistance at low temperatures. However, it also makes the unphysical prediction that the resistance will be infinite at even lower temperatures. It turns out that Kondos result is correct only above a certain temperature, which became known as the Kondo temperature, T_K.

The Kondo temperature is defined as the temperature at which the Kondo effect clearly appears and for which Kondos result is valid. It is given approximately by

$$T_K = E_F \sqrt{J} \exp(-1/nJ), \tag{3.72}$$

where T_K is the Kondo temperature, E_F is the Fermi energy, J characterizes the strength of the exchange interaction and n is the density of state. Generally T_K is lower than the resistance minimum T_M that can be estimated from the approximate expression giving the resistivity $R(T)$,

$$R(T) = a - b \ln T + cT^5. \tag{3.73}$$

The $\ln T$ term contains the spin-dependent Kondo scattering and cT^5 characterizes the resistivity due to phonon scattering at low temperature (the low temperature is also required for a sharp Fermi surface), and a, b and c are constants with b being proportional to the exchange interaction. This leads to a resistivity minimum at approximately

$$T_M = (b/5c)^{1/5}. \tag{3.74}$$

The theoretical framework for understanding the physics below T_K emerged in the late 1960s from Phil Andersons idea of "scaling" in the Kondo problem. Scaling assumes that the low temperature properties of a real system are adequately represented by a coarse-grained model [99]. As the temperature is lowered, the model becomes coarser and the number of degrees of freedom it contains is reduced. This approach can be used to predict the properties of a real system close to absolute zero. The Anderson impurity model and accompanying renormalization theory was an important contribution to understanding the underlying physics of the problem [100]. The resemblance of the theory with the Hubbard model is well described in [101] and the relation between the Kondo model and the Anderson model has also studied in [102].

3.5. Heisenberg Magnet

In this section, we will briefly overview the derivation of generalized Heisenberg Hamiltonian which is essential to the quantum mechanical theory of low temperature ferromagnetism. This Hamiltonian is often used to explain the properties of coupled spin systems and magnons. Magnon is a collective excitation of the electrons' spin structure in a crystal lattice [19]. As it was already seen in the simple case, the Heisenberg Hamiltonian is useful in showing how an electrostatic exchange interaction approximately predicts the existence of a molecular field and hence gives a fundamental qualitative explanation of the existence of ferromagnetism.

Beginning with two atoms having outmost free electrons, the two electron system cannot be excited thermally in the higher lying states of the two atoms so that the state can asymptotically be ignored. In the approximation, it is legitimate to consider only the lowest lying singlet and triplet states and simple effective Hamiltonian replaces the general Hamiltonian by determining the state of the electrons in the constraint subspace. It was pointed out by Heisenberg [120] and Dirac [121]. In that case, the effective Hamiltonian for the two electrons having electron spins $\vec{\sigma}_1$ and $\vec{\sigma}_2$ is proportional to

$$H_{\text{two}} = -J\left(\vec{\sigma}_1 + \vec{\sigma}_2\right)^2 + \text{constant} = -2J\vec{\sigma}_1 \cdot \vec{\sigma}_2 + \varepsilon_0, \tag{3.75}$$

where ε_0 is zero point energy and J is the exchange constant. The total spin operator identity $|\vec{\sigma_1}|^2 = |\vec{\sigma_2}|^2 = 3/4$ and the notation $\vec{\sigma}_i = (\sigma_i^x, \sigma_i^y, \sigma_i^z)$ with the site index i are used in the formula. This form of the effective Hamiltonian is known as Heisenberg model whose explicit expression has already seen in (3.62). When $J > 0$, the lowest energy state of the Hamiltonian has parallel spins whose magnetic character follows the ferromagnetism and, when $J < 0$, the state has antiparallel spins to be antiferromagnetism. The exact solution already has been discussed in the previous section.

Generalization of the Hamiltonian in an ionic lattice is also known as Heisenberg model. When there is especially, 1D lattice with nearest neighbor interaction, it is called spin chain model and sometimes called as XX model or XY model which depends upon the type of interaction.[j] When the coupled spins are exposed to uniform external magnetic field B, the general description of the 1D model becomes as

$$H_{1D} = -\sum_i J_i^x \sigma_i^x \sigma_{i+1}^x + J_i^y \sigma_i^y \sigma_{i+1}^y + J_i^z \sigma_i^z \sigma_{i+1}^z + \frac{B}{2} \sum_i \sigma_i^z. \tag{3.76}$$

With the uniform interaction strength in the XX and YY direction, $J_i^x = J_i^y = J$, the first two terms mean the electron exchange interactions, since $\sigma_i^x \sigma_{i+1}^x + \sigma_i^y \sigma_{i+1}^y = \sigma_i^+ \sigma_{i+1}^- + \sigma_i^- \sigma_{i+1}^+$ represents the electron hopping from the site i to $i+1$. Here σ_i^+ and σ_i^- are the spin creation and annihilation operators defined as $\sigma^\pm \equiv (\sigma_i^x \pm \sigma_i^y)/\sqrt{2}$. When $J_i^x = J_i^y = 0$, the model is identical to the 1D classical Ising model because the only possible matrix elements of σ^z are either $+1$ or -1 and they share same partition function. The Ising model has been solved exactly in 1D and 2D lattices and the results are well summarized in [122, 123].

The 1D Heisenberg model in (3.76) is not easy to handle in general but it is known that the model is exactly soluble in the various limits using well-known operator transformation techniques [116]. We will discuss the solutions in Chapter 4 as we discuss the pure quantum mechanical properties of the model.

3.6. Spinless Hubbard Model: Superfluidity

In order to remain close to the physical system, we consider the so-called Bose–Hubbard model. This model is used to investigate various superconducting and superfluid behaviors, ranging from high T_c superconductors to cold atom gases in optical lattices. In this section, we focus on the model describing superfluidity which was one of the

[j]When the interaction coupling strength of $\sigma_i^x \sigma_{i+1}^x$ and $\sigma_i^y \sigma_{i+1}^y$ is different, the model is called XY model or anisotropic spin chain. When the coupling strength is same, the model is called XX model or isotropic spin chain.

most surprising discovery in the mid-1990. It is a particular liquid phase whose properties cannot be explained in the regime of classical physics. An extensive historical overview of the superfluidity can be found at [124].

Superfluidity is a state of matter in which the matter behaves like a fluid without viscosity and with infinite thermal conductivity. The substance, which appears to be a normal liquid, will flow without friction, when they pass any surface, which allows it to continue to circulate over obstructions and through pores in containers which hold it, subject only to its own inertia. Since even gases have viscosity, superfluids have less resistance to shear than a gas does. The phenomena have been found in the molecule having simple structure-like liquid Helium at a low temperature (below 2.18 K called lambda point). Helium-4 atoms[k] are bosons and the superfluidity of helium-4 can be regarded as a consequence of Bose–Einstein condensation in an interacting system (Fig. 3.7).

The Bose–Hubbard model has quite a complex phase diagram [33], but we will only be interested in some special regimes relevant to our study of the model. The Bose–Hubbard Hamiltonian is obtained through the replacing the fermion operators by bosonic operators in the Hubbard model (3.49) as

$$H_{\mathrm{BH}} = J\sum_i (b_i^\dagger b_{i+1} + b_i b_{i+1}^\dagger) + \frac{U}{2}\sum_i (n_i - 1)n_i, \tag{3.77}$$

where the first term describes the nearest neighbor site hopping of bosons, ($b_i^\dagger$ and b_{i+1} are the usual bosonic raising and lowering operators, respectively), and the second term is the on-site repulsion between bosons. In the limit of low density, the n_i^2 term can be ignored and we obtain the so-called XX Hamiltonian:

$$H = -J\sum_i \sigma_i^x\sigma_{i+1}^x + \sigma_i^y\sigma_{i+1}^y - \mu\sum_i \sigma_i^z, \tag{3.78}$$

where $\sigma^x = b + b^\dagger$, $\sigma^y = i(b - b^\dagger)$ and $\sigma^z = 1 - 2b^\dagger b$ are the usual Pauli matrices and we can think of μ as the chemical potential [28]. They represent

[k]Helium atom has two different types of isotope, helium-3 and helium-4. Though the phenomena of the superfluid states of helium-4 and helium-3 are very similar, the microscopic details of the transitions are very different. On the other hand, helium-3 atoms are fermions, and the superfluid transition in this system is described by a generalization of the BCS theory of superconductivity. In the superconductivity, cooper pairing takes place between atoms rather than electrons, and the attractive interaction between them is mediated by spin fluctuations rather than phonons.

Fig. 3.7. The image of superfluidity. Liquid helium at a low temperature becomes superfluid phase and all the molecules in the material behave as like as a single object. Without any force applied, the liquid climbs up the container's wall and is dripped off at the opposite-bottom side of the container. The image has been taken from the Google image section.

two-level systems where the states $|0\rangle$ and $|1\rangle$ are the boson occupation numbers at each site (an empty site and one boson in the site respectively [34]). Physically this should be clear, since by making the density very low, we preclude more than one particle from occupying each of the sites. As we have been seen, the Hamiltonian would also be obtained if we investigate the spinless fermion Hubbard model since fermions obey the Pauli exclusion principle and therefore we cannot have more than one electron per site.

The XX model has been extensively studied and its spectrum is well understood through the Jordan–Wigner transformation [28]. However, since we would like to understand superfluidity in the original Hubbard model we need to know the response of the system with such a Hamiltonian to introducing external perturbations (here we study a 1D model; two dimensions are needed for proper superfluidity, but the 1D discussion leads to the same conclusion — see Appendix A.2 for a discussion of the point). In order to study criticality in the XX model, we introduce the so-called "twisted" Hamiltonian [35, 36]:

$$H_\theta = -J\sum_i e^{i\theta}\sigma_i^+\sigma_{i+1}^- + \sigma_i^-\sigma_{i+1}^+ e^{-i\theta} - \mu\sum_i \sigma_i^z \tag{3.79}$$

obtained by imposing Peierls phases on each Pauli raising and lowering operator: $\sigma^- \to e^{i\theta}\sigma^-$ and $\sigma^+ = e^{-i\theta}\sigma^+$. We can think of phases arising from an imposition of an external field (like, for example, the Aharonov–Bohm effect, where the charge encirculating the vector potential, A, gains the phase $e^{-i\int Adl}$). The twisted Hamiltonian therefore represents the response of the system to an external disturbance and this is what gives rise to the physics of various critical phenomena.

The superfluid density can now be obtained from the assumption that the difference between the normal and "twisted" energy is — to the lowest order — given by the superfluid kinetic energy. So, when we externally weakly perturb the system, its response is to move its superfluid component (if there is such a component). If the response is null to the lowest order, our material is not in its critical phase. Translated into mathematics, our statement reads [37]:

$$\langle H_\theta \rangle - \langle H \rangle = \frac{1}{2} N f_s m v_s^2, \tag{3.80}$$

where N is the number of particles in the superfluid and m is the mass of each particle so that mNf_s is total mass of superfluid component. v_s is the superfluid velocity. The superfluid density is denoted by f_s. The superfluid velocity is given by $v_s = \frac{\hbar}{m}\nabla\theta$, where θ is the phase of the superfluid macroscopic wave function — the order parameter (and is the same angle that enters the twisted Hamiltonian). We will assume that the superfluid phase varies linearly with distance, $\theta(x) = \theta_T x/L$. From this definition, it is clear that the superfluid fraction, f_s, can be expressed as

$$f_s = \frac{2m}{\hbar^2}\frac{L^2}{N}\frac{\langle H_\theta \rangle - \langle H \rangle}{\theta_T^2} = \frac{1}{JN}\frac{\langle H_\theta \rangle - \langle H \rangle}{\theta^2}, \tag{3.81}$$

which is essentially the energy difference between the perturbed and the original Hamiltonian, divided by the phase squared. Note that here we used the fact that $J = \hbar^2/2ma^2$ and $\theta = \theta_T/N$, where a is the typical lattice spacing.

Assuming the θ is very small, $H_\theta - H$ can be computed up to second-order term. The phase factor can be expanded as $e^{i\theta} \simeq 1 + i\theta + (i\theta)^2/2! + \cdots$. Computing $\langle H_\theta - H \rangle$ to the second order of θ, we obtain the superfluid fraction XX to be:

$$f_s = \frac{1}{2N}\left\langle \sum_i \sigma_i^x \sigma_{i+1}^x + \sigma_i^y \sigma_{i+1}^y \right\rangle - \frac{i}{N\theta}\left\langle \sum_i \sigma_i^+ \sigma_{i+1}^- - \sigma_i^- \sigma_{i+1}^+ \right\rangle_2 . \tag{3.82}$$

The second term, usually known as the superfluid current [36], vanishes for our translationally invariant Hamiltonian in the Bogoliubov limit. It describes the electron currents because it is difference of the electron hopping probability. The subscript "2" indicates that the average is a more complicated second-order one, but the exact details beyond out scope and can be found at [36]. The condition for non-zero superfluid density, $f_s > 0$, now leads to

$$\left\langle \sum_i \sigma_i^x \sigma_{i+1}^x + \sigma_i^y \sigma_{i+1}^y \right\rangle \neq 0. \tag{3.83}$$

So the supercurrent and therefore criticality exist as long as the above expectation remains finite, which in turn implies that the condition for criticality is $\mu < J$ (so long as we are at $T = 0$). When, on the other hand, $\mu \geq J$, the superfluid density identically vanishes. The critical point of the model is therefore $\mu = J$ and the state above critical point is $|00...0\rangle$, signifying that all sites in the system are empty. Physically this is because the chemical potential, μ, is too high for any particles to exist (note that the μ influences the form of the ground state and, therefore, it also affects the expectation value determining the superfluid fraction, even though it does not appear explicitly in Eq. (3.83)).

3.7. Superconductivity

Another important low temperature phenomenon in solid is superconductivity. At a low temperature, the electric resistivity in a certain material becomes suddenly disappeared and it becomes superconductor.

The resistance in a material has two component, due to (i) thermal vibrations of the atoms and (ii) impurities and defects in the crystal. Both these processes cause the scattering of electrons while electrons are not scattered by a perfect lattice. As the temperature is decreased, the lattice vibrations decrease, and it is consequently observed that the resistivity of the material is also reduced in a continuous way. Contrary to such behavior, the change of resistivity in superconducting material is abrupt and becomes zero suddenly before the temperature reached to absolute zero. The behavior is sketched in Fig. 3.8.

The first discovery of the phenomena had been made by H. Kamerlingh Onnes in 1911 who measured the electrical resistivity of mercury and found that it dropped to zero below 4.15 K. He was able to do the

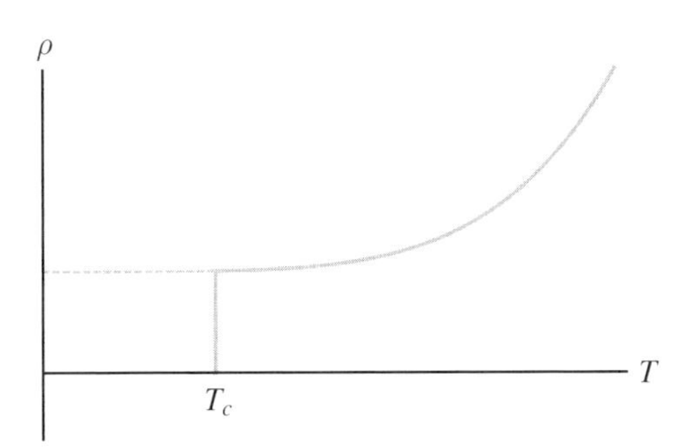

Fig. 3.8. Change of electric resistivity ρ in the superconducting material at a low temperature. The material changes its phase below the critical temperature and turns into a superconductor.

experiment because he was the first to liquefy helium. It allows him to work with the low temperatures required for superconductivity. After his discovery, many experiment had been performed and the superconducting phenomena in the other materials had been observed even at a higher temperature.

In 1933, Meissner and Ochsenfeld made another fundamental discovery on the superconducting material [126]. They found that superconductors expelled magnetic flux when they were cooled below the transition temperature. This establishes that there was more to the superconducting state than just perfect conductivity (which would require $E = 0$); it is also a state of perfect diamagnetism or $B = 0$ inside of material when there is external magnetic field if it is in the superconducting phase.

Since the first discovery of superconductivity by Onnes, it took 46 years before Bardeen, Cooper, and Schrieffer (BCS, 1957) presented a theory that correctly accounted for a large number of experiments on superconductors. Even today, the theory of superconductivity is rather intricate and so perhaps it is best to start with a qualitative discussion of the experimental properties of superconductors.

The superconductive state is a macroscopic state. This has led to the development of superconductive quantum interference devices that can be used to measure very weak magnetic fields. We will briefly discuss this after we have laid the foundation on tunneling involving superconductors. We will then discuss the BCS theory and show how the electron–phonon interaction can give rise to an energy gap and a coherent motion of electrons without resistance at sufficiently low temperatures.

3.7.1. *Basic explanation of superconductivity: Brief history*

Theory on the emergence of the superconductivity begins by explaining the effect of zero resistances. Electron in a lattice is scattered by imperfection of the lattice (mainly by lattice vibration and the impurity of the lattices) so that the resistance is the function of temperature and homogeneousness of the material. On the contrary, the superconductivity is a phenomenon which is actually benefited from the lattice vibration. It is argued that the electric current can be enhanced due to the interaction between the electrons and the quantized vibration mode in the lattices at a quantum mechanical regime.

Such a possibility had firstly suspected by German born British scientist whose conjecture was supported by the experimental evidence. In 1950, Frohlich [127] discussed the electron–phonon[1] interaction and considered the possibility that this interaction might be responsible for the formation of the superconducting state. At about the same time, Reynolds, Serin, Wright, and Nesbitt [125] found that the superconducting transition temperature depends on the isotopic mass of the atoms of the superconductor. They found that the critical temperature scales inversely to the mass of atoms in the lattice

$$T_c \sim \frac{1}{M^\alpha} \tag{3.84}$$

where α is some power. This experimental result gave strong support to the idea that the electron–phonon interaction was involved in the superconducting transition. It is because the atomic vibration (whose frequency depends on M) indicates electron attraction. In the simplest model, $\alpha = 1/2$.

In 1957, Bardeen, Cooper, and Schrieffer (BCS) finally developed a formalism that contained the correct explanation of the superconducting state in common superconductors [129]. Their ideas had some similarity to Frohlichs. The key idea of the BCS theory was developed by Cooper in 1956 [128]. Cooper analyzed the electron–phonon interaction in a different way from Frohlich. Frohlich had discussed the effect of the lattice vibrations on the self-energy of the electrons. Cooper analyzed the effect of lattice vibrations on the effective interaction between electrons and showed that an attractive interaction between electrons (even a very weak attractive interaction at low enough temperature) would cause pairs of the electrons

[1]Phonon is quantized particle of lattice vibrational mode.

(the Cooper pairs) to form bound states near the Fermi energy. Later, the theory has been developed by BCS and they were able to describe a cooperative condensation process in which many pairs of electrons of the normal Fermi sea are formed so to minimize the total energy of the system. We will discuss the BCS theory and show the pairing interaction causes a gap in the density of single-electron states.

3.7.2. *Electron–electron interaction — BCS model*

The basic idea of BCS theory is the reduction of resistivity through the attraction between electrons having opposite momentum located in a lattice. An electron passing through a crystal lattice can cause a transient distortion of the positive lattice ions as it passes them. This vibrational distortion of the lattice can then attract another electron. The distortion-induced attraction is opposed by the short-range Coulomb repulsion between the two negative electrons. Broadly speaking, conventional superconductivity occurs in those materials for which the transient, lattice-mediated attraction is stronger than the Coulomb repulsion.

The Coulomb potential between two electrons is inversely proportional to their distance. The more they are separated the less their repulsive force is. However, if the two electrons are located inside atomic structures or periodic atomic lattices, the Coulomb force are screened by the atoms in the between and the strength of the force is dropped off exponentially faster as its distance is increased. In addition, the extra force between the electrons will be generated due to the interaction between the electrons and the lattices.

We start our discussion from the interaction between the free electron and the lattice phonon. In the electron–phonon interaction, it can be assumed that the electron–electron interaction are screened by the ions and the ions interacts with each other only through the electrons which are subjected to a short-range screened potential. In that case, the electrons themselves can be treated as independent fermions. For a Bravais lattice the unperturbed Hamiltonian is simply

$$H_0 = \sum_k \epsilon_k c_k^\dagger c_k + \sum_{q,s} \hbar\omega_{qs} a_{qs}^\dagger a_{qs}, \tag{3.85}$$

the phonon frequencies ω_{qs} being proportional to q as q goes zero. To this we add the interaction between the electrons and the screened ions. We assume that, at any point r, the potential due to a particular ion depends only on the equilibrium position of the lth ion, r_l, and the distance of the ion's displacement x_l so that in second-quantized notation of the interaction Hamiltonian

$$H_I = \sum_{k,k',l} \langle k|V(r_l)|k'\rangle c_k^\dagger c_{k'} = \sum_{k,k',l} e^{i(k-k')\cdot(l+x_l)} V_{k-k'} c_k^\dagger c_{k'}, \tag{3.86}$$

where l is the equilibrium position of the lth ion.[m] Here $V(r_l)$ is the potential due to a single ion at the origin r_l and $V_{k-k'}$ is its Fourier transformed potential. With the assumption that the displacement of the ion is sufficiently small that $(k-k')\cdot x_l \ll 1$ then the exponential part can be expanded as $e^{i(k-k')\cdot x_l} \simeq 1 + i(k-k')\cdot x_l + \cdots$ and it splits the interaction Hamiltonian into two parts,

$$H_I = H_{\text{Bloch}} + H_{\text{el-ph}} \tag{3.87}$$

where the first term H_{Bloch} represents the electron scattering independent to the lattice displacement and the second term is the electron–phonon interaction term. They are

$$H_{\text{Bloch}} = \sum_{k,k',l} e^{i(k-k')\cdot l} V_{k-k'} c_k^\dagger c_{k'} = N \sum_{k,g} V_{-g} c_{k-g}^\dagger c_k, \tag{3.88}$$

$$H_{\text{el-ph}} = i\sqrt{N} \sum_{k,k'} (k-k')\cdot x_{k-k'} V_{k-k'} c_k^\dagger c_{k'}, \tag{3.89}$$

where g is the reciprocal lattice vectors and we have used the relation $x_l = N^{-1/2} \sum_q e^{iq\cdot l} x_q$. In terms of the annihilation and creation operators defining canonical operator as $x_l = \sqrt{\frac{\hbar}{2M\omega_l}}(a_l^\dagger + a_l)$, the electron–phonon interaction hamiltonian becomes,

$$H_{el-ph} = i \sum_{k,k'} \left(\frac{N\hbar}{2M\omega_{k-k'}}\right)^{1/2} (k-k') V_{k-k'} (a_{k'-k}^\dagger + a_{k-k'}) c_k^\dagger c_{k'}, \tag{3.90}$$

where we assume the phonon spectrum to be isotropic[n] and the phonon has the longitudinal modes only, for which it is parallel to $k'-k$. In addition, we shall neglect the effects of H_{Bloch}, due to the periodic potential condition

[m] One can think r_l is the equilibrium position of the ion with unit of l as $r_l = 2\pi l/a$ with the lattice spacing a.

[n] Because x_l and x_{-l} represent the same phonon mode, extra care should be taken in using the expression for the hamiltonian of electron–photon interaction.

of the stationary lattice, $V_x = V_{q+x}$. With these simplifications, we are left with the Fröhlich Hamiltonian,

$$H = \sum_k \epsilon_k c_k^\dagger c_k + \sum_q \hbar\omega_q a_q^\dagger a_q + \sum_{k,k'} M_{k,k'}(a_{-q}^\dagger + a_q)c_k^\dagger c_k, \tag{3.91}$$

where the electron–phonon matrix element is defined by

$$M_{k,k'} = i\sqrt{\frac{N\hbar}{2M\omega_q}}(k' - k)V_{k-k'} \tag{3.92}$$

with the phonon wavenumber q equal to $k-k'$, reduced to the first Brillouin zone if necessary.

The interaction Hamiltonian can be considered as being composed of two parts: (a) the term involving $a_{-q}^\dagger c_k^\dagger c_{k'}$ and (b) terms involving $a_q c_k^\dagger c_{k'}$. They represent the electron scattering processes (a) from k' to k with the emission of a phonon of wavenumber $k' - k$ and (b) from k' to k with the absorption of a phonon of wavenumber $k-k'$. The total wavenumber is then conserved, as is always the case in the periodic system, unless the vector $k' - k$ lies outside the first Brillouin zone, so that $q = k' - k + g$ for some non-zero g. In that case, such electron–phonon Umklapp processes do not conserve wavenumber and are important in contribution to the electrical resistivity of metals.

Since these lattice deformation are resisted by the same stiffness that makes a solid elastic, it is clear that the characteristic vibrational, or phonon, frequencies will play a role. For the electronically screened Coulomb interaction, the characteristic frequency is the plasma frequency, which is so high that we can see that if an electron is scattered from k to k', the relevant phonon must carry the momentum $q = k-k'$ and the characteristic frequency must then be the phonon frequency ω_q. As a result, it is plausible that the phonon contribution to the screening function be proportional to $(\omega^2 - \omega_q^2)^{-1}$.

Electrons deform the lattice (as they distract ions in the lattice) and this deformation acts on other electrons in a way that they are attracted each other. The mechanism leading to the indirect electron–electron interaction via phonons is schematically represented in Fig. 3.9: one electron of wave vector k_1 emits a phonon of wave vector q (vibration mode of the lattice) and is scattered into the state $k_1 - q$; the phonon of wave vector q is immediately "absorbed" by another electron wave vector k_2 that scatters into the state $k_2 + q$. Under the energy conservation, the total

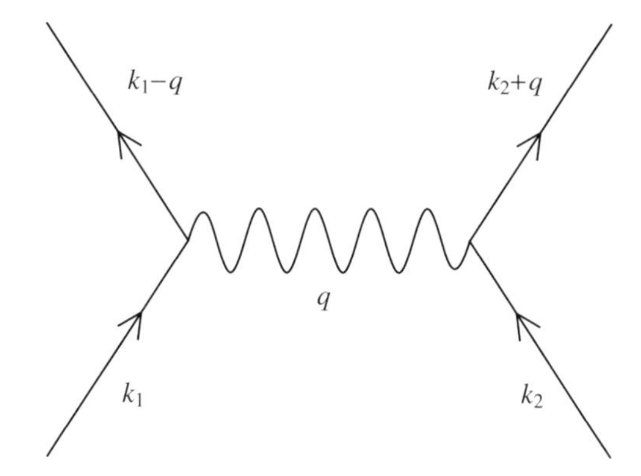

Fig. 3.9. The virtual exchange of a phonon of wave vector q. The k are the wave vectors of the electrons. The diagram illustrates the effective electron–electron interaction which is caused by the electron–phonon interaction. This is the fundamental process of superconductivity.

energy is conserved and electrons exchange their momentum as $k_1 - k_2 = q$ (Fig. 3.9).[o] The quantum mechanical analysis of this process provides a note attractive interaction between pairs of electrons, whose energy difference is smaller than the phonon energy. This leads to an effective coupling between electrons (see Appendix A.3 for the derivation) and they are

$$H_{\text{BCS}} = \sum_{k,\sigma} \epsilon_k c^\dagger_{k\sigma} c_{k\sigma} + \sum_{k_1,k_2,q,\sigma_1,\sigma_2} W_{k_1,k_2,q} (c^\dagger_{k_2+q,\sigma_2} c^\dagger_{k_1-q,\sigma_1} c_{k_1,\sigma_1} c_{k_2,\sigma_2}), \tag{3.93}$$

where the effective scattering coefficient has the form as

$$W_{k_1,k_2,q} = \frac{N(\hbar q V_q)^2}{2M[(\epsilon_{k_1} - \epsilon_{k_1-q})^2 - (\hbar\omega_q)^2]}. \tag{3.94}$$

The spin σ_i of the ith electron has also been explicitly included in this transformed Hamiltonian.

The ground state of the BCS Hamiltonian equation (3.93) can be found rather easily with further assumption. Consider the superconducting electron system as being in some sort of condensed phase and it thus becomes reasonable to make the hypothesis that, in the superconducting problem, the scattering of pairs of electrons having equal but opposite

[o]For the perfect energy transfer or elastic scattering, the condition is $k_1 = k_2 + q$ or equivalently $k_2 = k_1 - q$.

momentum will be similarly important. We refer to these as Cooper pairs and accordingly retain from the interaction only those term for which $k_2 = -k_1 \equiv k$ and $k_2 + q = -k_1 + q \equiv k'$. In that case, the reduced Hamiltonian of the paired electrons becomes

$$H_{\mathrm{BCS}}^{k_2=-k_1} = \sum_{k,\sigma} \epsilon_k c_{k,\sigma}^\dagger c_{k,\sigma} - \frac{1}{2} \sum_{k,k',\sigma_1,\sigma_2} V_{k,k'} (c_{k',\sigma_2}^\dagger c_{-k',\sigma_1}^\dagger c_{-k,\sigma_1} c_{k,\sigma_2}), \tag{3.95}$$

where in the notation $V_{k,k'} = -2W_{-k,k,k'-k} - U_{k,k'}$ with $U_{k,k'}$ a screened Coulomb repulsion term which we add to the Fröhlich Hamiltonian. A positive value of $V_{k,k'}$ thus corresponds to a net attractive interaction between electrons.

In the Hamiltonian, the spin degree of freedom can be further simplified. The spin index σ can have both values $\uparrow$ and $\downarrow$. If we use the condition for the potential symmetrization, $V_{k,k'} = V_{-k,-k'}$, the notation can be further abbreviated by adopting the convention that an operator written with an explicit minus sign in the subscript refers to spin $\downarrow$ while an operator without a minus sign refers to $\uparrow$, e.g. $c_k^\dagger \equiv c_{k,\uparrow}^\dagger$, $c_{-k}^\dagger \equiv c_{-k,\downarrow}^\dagger$ and so on. Then, we have a simplified effective BCS Hamiltonian,

$$H_{\mathrm{BCS}} = \sum_k \epsilon_k (c_k^\dagger c_k + c_{-k}^\dagger c_{-k}) - \sum_{k,k'} V_{k,k'} (c_{k'}^\dagger c_{-k'}^\dagger c_{-k} c_k), \tag{3.96}$$

where $c_k^\dagger$ is the creation operator of the spinless Fermi particle. This is the Hamiltonian of the BCS model of which the eigenstates and eigenvalues are going to be explored.

The solution of the BCS Hamiltonian can be obtained from the standard diagonalization technique developed by Bogoliubov which has been used to solve the Hubbard model (see for example [130]). Instead of solving the Hamiltonian directly, we approach the problem by presenting the eigenstate of the Hamiltonian which is corresponding to the state of electron–electron pairs in the momentum state. Such a state is to be generated by placing the paired electrons (Cooper pair) having opposite momentum in the vacuum state as

$$|\psi_{\mathrm{BCS}}\rangle = \prod_k (u_k + v_k c_k^\dagger c_{-k}^\dagger)|0\rangle, \tag{3.97}$$

where $|0\rangle$ is empty state and k is the index of the electron momenta. The u_k and v_k are arbitrary complex numbers whose values are determined by the variables in the Hamiltonian.

Since the Hamiltonian and the eigenstate exist in the Hilbert space of the pair-wised Fermi particles, all of the operators can be projected into two-level spin system in SU(2) space (pseudo spin space). It is because the possible paired fermion states are only $|0_k, 0_{-k}\rangle \equiv |\downarrow\rangle$ or $|1_k, 1_{-k}\rangle \equiv |\uparrow\rangle$ such that they exist in the 2D Hilbert space. The mapping also exploit the fact that the paired fermion is invariant under the particle exchange with the other pairs which are all taken as an independent single particle excitation. Creation of the paired fermion $c_k^\dagger c_{-k}^\dagger$ is equivalent to the flipping the 1/2-spin (from the down-spin to up-spin) while the energy of the created pair $c_k^\dagger c_k + c_{-k}^\dagger c_{-k}$ corresponds to the energy difference between the two spin directions (up-spin and down-spin). A formal expression of the operator equivalence can be given by

$$\sigma_k^z = c_k^\dagger c_k + c_{-k}^\dagger c_{-k} - 1, \tag{3.98}$$

$$\sigma_k^\dagger = c_k^\dagger c_{-k}^\dagger, \tag{3.99}$$

which is quite useful in the treatment of BCS Hamiltonian.

Using the mapping, the BCS Hamiltonian equation (3.96) is now safely modified into a reduced spin Hamiltonian (the approach we follow is due to Anderson) which is the Heisenberg model at certain limit. When the scattering coefficient is uniform for the different modes as $V_{k,k'} \equiv V$, the reduced BCS Hamiltonian becomes

$$H_{\text{BCS-Reduced}} = \sum_k \epsilon_k(\sigma_k^z + 1) - \frac{V}{2}\sum_{kk'}(\sigma_{k'}^x \sigma_k^x + \sigma_{k'}^y \sigma_k^y) \tag{3.100}$$

which is identical to the specific limit of Heisenberg interaction. As we have discussed before, the Hamiltonian can be diagonalized exactly and one can analyze its ground state with the solution.

From the model, it is possible to identify the critical temperature where the system become superconductor. If we use the mean-field approach, the interaction part of the Hamiltonian equation (3.100) can be approximated by a single spin operator in a rotated direction. The kth mode component of the mean-field Hamiltonian is

$$H_{\text{MF}}^k = \epsilon_k \sigma_k^z - \frac{V}{2}(S_x \sigma_k^x + S_y \sigma_k^y)$$

where S_x and S_y are the spin expectation value as $S_x = \sum_k \langle \sigma_k^x \rangle$ and $S_y = \sum_k \langle \sigma_k^y \rangle$ where $\langle \sigma_k^x \rangle = \mathrm{Tr}[\sigma_k^x \rho]$ and $\rho = e^{-\beta H_{\mathrm{MF}}^k}/Z$. We expect the critical temperature for the phase transition roughly below the interaction strength. It can be identified by self-consistency equation which is established as

$$S_x = \sum_k \langle \sigma_k^x \rangle \implies \sum_k V \left(\frac{\tanh \beta \sqrt{\epsilon_k^2 + \Delta^2}}{\sqrt{\epsilon_k^2 + \Delta^2}} \right) = 1, \tag{3.101}$$

where $\Delta = V\sqrt{S_x^2 + S_y^2}/2$. The energy gap parameter Δ is the part of the eigenvalues for the mean-field Hamiltonian, $\pm\sqrt{\epsilon_k^2 + \Delta^2}$. The value Δ determines the ground state energy and it is proportional to the interaction strength and mean-value of the magnetic field. Assuming that the energy density ρ_f of the fermion is uniformly distributed over k, the summation in the self-consistency equation become integral and Eq. (3.101) is reduced into

$$V\rho_f \int_0^{\hbar\omega_D} \frac{\tanh\left[\beta\sqrt{\epsilon^2 + \Delta^2}\right]}{\sqrt{\epsilon^2 + \Delta^2}} d\epsilon = 1, \tag{3.102}$$

where $\hbar\omega_D$ is the Debye energy. At zero temperature $T = 0$ this equation for $\Delta_{T=0}$ reduces to

$$\Delta_{T=0} = 2\hbar\omega_D \exp\left[-1/V\rho_f\right], \quad \text{when } T = 0. \tag{3.103}$$

Given that the typical Debye energy is 0.03 eV and $V\rho_f$ is the order of 0.1, the $\Delta_{T=0}$ is a very small quantity indeed, being generally about one percent of the Debye energy and hence corresponding to thermal energies at temperature of the order of 1 K. When the temperature is raised above zero the numerator of the integrand in Eq. (3.102) is decreased and so in order for the equation is to be satisfied the denominator must be also decreased. This implies that Δ is a monotonically decreasing function of T. In fact, it has the form shown qualitatively in Fig. 3.10.

The Δ drops as the temperature increased and disappear at certain critical temperature T_c. The critical temperature can be obtained from the self-consistent equation by setting $\Delta = 0$ which becomes

$$V\rho_f \int_0^{\hbar\omega_D} \frac{\tanh\left[\beta_c \epsilon\right]}{\epsilon} d\epsilon = 1, \tag{3.104}$$

where β_c gives the critical temperature $\beta_c = 1/k_B T_c$. The magnitude of the transition temperature T_c in the BCS model is found by evaluating the

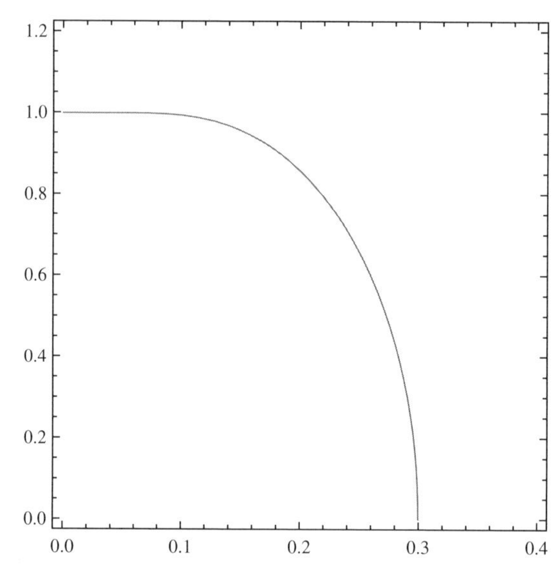

Fig. 3.10. The energy gap parameter Δ decreases as the temperature is raised from zero and vanishes at the transition temperature T_c.

integral whose solution is obtained as (see Appendix A.4)

$$k_B T_c = 1.14\ \hbar\omega_D e^{-\frac{1}{\rho_f V}} \quad \text{or} \quad 2\Delta_{T=0} = 3.5\ k_B T_c.$$

The result is adequately agreed with the experimentally observed values. It has been known that the most of elements have the value $2\Delta/T_c$ between two and five.

3.7.3. *Meissner effect — Superconductor in magnetic field*

Besides the effect of zero resistivity below a critical temperature, Onnes further discovered that a superconducting state could be destroyed by placing the superconductor in a large enough magnetic field. He also noted that sending a large enough current through the superconductor would destroy the superconducting state. Silsbee later suggested that these two phenomena were related. The disruption of the superconductive state is caused by the magnetic field produced by the current at the surface of the wire. In 1933, Masissner and Oschsen Feld found that the superconduction material in magnetic field cancels magnetic flux inside of it and repel to the magnetic field. Differently from a perfect conductor, the superconductor push the magnetic force out of the body and it can be levitated against gravity if it is placed in weak magnetic field. The effect is called Meissner effect.

If one imagine a superconducting wire in a loop shape, a current generated in the loop will never be stopped due to the zero resistivity. The change of magnetic field in the material causes electromotive force reversed the influence of the magnetic field and the generated current in the superconducting material would last without interruption. However, if the magnetic field is strong enough, it can create the current to overcome the effect and the material turns into a normal conductor. In that sense, the state of superconductor is completely different phase with respect to normal conductor.

Technically speaking, it implies that superconductor behaves like a perfect diamagnetic material. Inside the superconductor, the magnetic field B is given as

$$B_{\text{super}} = \mu_0(H_{\text{ext}} + M) = 0 \tag{3.105}$$

where M is the magnetization due to the induced surface currents and H_{ext} the externally applied field, so that $M = -H_{\text{ext}}$ and $\chi = M/H = -1$. Exclusion of the flux is due to perfect diamagnetism caused by surface currents, which are always induced so as to shield the interior from external magnetic fields. A simple application of Faradays law for a perfect conductor would lead to a constant flux rather than excluded flux. The equation describing the critical fields dependence on temperature is often empirically found to obey

$$H_c(T) = H_0\left[1 - \left(\frac{T}{T_c}\right)^2\right], \tag{3.106}$$

where H_0 is the critical magnetic field to destroy the superconductivity at the zero temperature. Such a behavior of the critical magnetic field asymptotically, follows the change of energy gap parameter $\Delta(T)$ which is plotted in Fig. 3.10.

There are two different types of superconductivity depend upon the response behavior of the magnetic field. They are called as type I and type II. The superconductor mentioned above is the type I superconductor while the type II superconductor is the one which has two different values of critical magnetic field H_{c1} and H_{c2}. The behavior of the induced magnetic field M in type II superconductor follows the same trend in Eq. (3.105) upto H_{c1} and it is dropped exponentially to 0 at H_{c2}.

The critical induced current that destroys superconductivity is very structure sensitive so that it can be regarded for some purposes as an independent parameter. The critical magnetic field (that destroys superconductivity) and the critical temperature (at which the material becomes superconducting) are related in the sense that the highest transition temperature occurs when there is no external magnetic field with the transition temperature decreasing as the field increases.

To investigate further details of the superconductor's response to magnetic field, we study the theory of magnetic field in an superconducting material. In isolated atoms, the electronic wave function is rigid and unchanged to first order in an applied magnetic field, B. We will see below that the same must also be true for superconducting electrons. The total momentum p is therefore conserved in an applied magnetic field. The average electron velocity must be zero when the applied vector potential $A = 0$. The conservation of total momentum then requires $mv + qA = 0$ so that for an electron of mass m_e and charge $-e$, we can write $v = eA/m_e$ in an applied vector potential A. The resulting induced current density j is then given by

$$j = \rho(-e)v = -\frac{\rho e^2}{m_e} A, \tag{3.107}$$

where ρ is the density of electrons per unit volume. By assuming that we can associate a rigid (macroscopic) wave function with the ρ_s superconducting electrons per unit volume in a superconductor we derive that

$$j = \frac{-\rho_s e^2}{m_e} A. \tag{3.108}$$

Taking the curl of both sides we find that the magnetic field $B = \nabla \times A$ and current density j are related in a superconductor by

$$\nabla \times j = -\frac{\rho_s e^2}{m_e} B. \tag{3.109}$$

This relation was deduced from the Meissner effect by Fritz and Heinz London in 1935 and is referred to as the London equation. We have shown here how the London equation follows from the assumption of a rigid wave function and will show below how it leads to the Meissner effect.

To determine the variation of the magnetic field, B, inside a superconductor we can combine the London equation with Maxwell's steady-state equations for the magnetic field

$$\nabla \times B = \mu_0 j, \quad \nabla \cdot B = 0. \tag{3.110}$$

Taking the curl of the first of these Maxwell equation, we find

$$\nabla \times (\nabla \times B) = \mu_0 \nabla \times j. \tag{3.111}$$

If we use the London equation to replace $\nabla \times j$ on the right-hand side of the equation above, we can obtain the differential equation only with B. With the operator identity $\nabla \times \nabla \times B = \nabla(\nabla \cdot B) - \nabla^2 B$, we have second-order differential equation for the magnetic field

$$\nabla(\nabla \cdot B) - \nabla^2 B = -\frac{\mu_0 \rho e^2}{m_e} B \tag{3.112}$$

where ρ is the density of electrons per unit volume. Together with the second Maxwell's equation, the second-order differential equation describing magnetic field

$$\nabla^2 B = \lambda^{-2} B, \tag{3.113}$$

where $\lambda \equiv \sqrt{\frac{m_e}{\mu_0 \rho e^2}}$ is referred to as the London penetration depth. The Meissner effect follows immediately from the equation. After the consideration of the boundary condition, the 1D solution of the differential equation is given by

$$B(x) = B(0)\mathrm{e}^{-x/\lambda} \tag{3.114}$$

which describes the exponential decay of the magnetic field at the position x from the surface inside superconductor. The equation predicts the exponential decay of a magnetic field B away from the suffice of a superconductor and it became zero. λ characterizes the depth that the magnetic field can penetrate into and it is the only function of electron density.

The exponential decay of magnetic field is caused by the induced current at the surface of superconductor in the way as it is stated in the London equation. The strength of the current can be also found from the induced field by electrons. To understand the Meissner effect based upon the BCS theory, quantum mechanical model of the induced current should be derived. It is enough that the microscopic model of electric current is consistent with the London equation and the detailed analysis of it has been summarized in Appendix A.5. Quantum mechanically, the second quantized current is given as

$$j = \frac{e}{2mi}\left(\psi^\dagger \nabla \psi - \psi \nabla \psi^\dagger\right) - \frac{e^2}{mc}\psi^\dagger A \psi, \tag{3.115}$$

where the coefficient of the second term comes from the velocity of an electron in a vector potential $\vec{v} = \frac{1}{m}[p - \frac{e}{c}A]$. It can be shown that the first part vanishes for a perturbed BCS state (see Appendix A.6).

$$j = -\frac{e^2}{mc}\langle \psi_{\text{BCS}}|\psi^\dagger A \psi|\psi_{\text{BCS}}\rangle \tag{3.116}$$

$$= -\frac{e^2}{mc}A\underbrace{\langle \psi_{BCS}|\psi^\dagger \psi|\psi_{BCS}\rangle}_{n\text{-density}} \tag{3.117}$$

which is London equation stating that the electron current density is proportional to the function of the electron density. This tells us that the electric current is the function of the electron density and the applied field which had been stated in London equation (3.109).

However, differently from the normal conductor, the mechanism how the superconductor excludes magnetic field inside the material is still remained unanswered. As it is mentioned in the BCS theory, the ground state of superconducting Hamiltonian requires the paired electrons to exist in the each momentum space. In fact, they are the cause of increasing electron densities and repelling the magnetic field inside the material. The paired electrons are bound to the periodic potential which can be modeled by lattices. The state of one Cooper-pair superconductor becomes

$$(c^\dagger_{1\uparrow}c^\dagger_{1\downarrow} + c^\dagger_{2\uparrow}c^\dagger_{2\downarrow} + \cdots c^\dagger_{N\uparrow}c^\dagger_{N\downarrow})|0\rangle \tag{3.118}$$

This state exhibits long-range correlations. The paired electrons in the periodic potential is sketched in Fig. 3.11. If we find a Cooper pair in site 1, there is a high chance that we would not find any in site N and

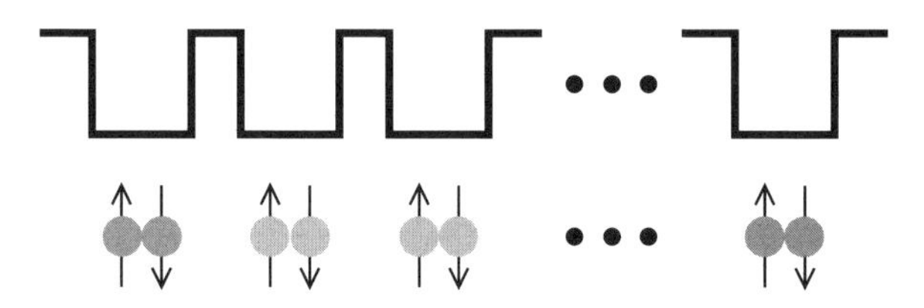

Fig. 3.11. Schematic diagram for the Cooper pairs in lattice sites.

vice versa. Correlation of this type is crucial for the superconductivity and superfluidity and the concept was firstly introduced by Penrose, Onsager and Yang [23] in 1962 under the name of off diagonal long-range order.

Suppose that we exchange two electron pairs, one at the site 1 and one at the site N. We can imagine doing this adiabatically, although the requirement of adiabaticity is by no means necessary (it is merely convenient, as this evolution then generates no other effect apart from the one we wish to concentrate on). Suppose that whatever the total state is, the reduced density matrix of sites 1 and N has a non-vanishing component of the state

$$|\Psi\rangle = |0_1\rangle|1_N\rangle + |1_1\rangle|0_N\rangle, \tag{3.119}$$

which, from the above discussion, means that the system has long-range correlation [23]. Then, after the swap, this component will look like

$$|\Psi\rangle = |0_1\rangle|1_N\rangle + e^{i\Phi}|1_1\rangle|0_N\rangle, \tag{3.120}$$

where $\Phi = \int Adl$ is the line integral of the vector potential along the path traversed by the electron pair (with proper units introduced below). The complete path around the magnetic field made by the Cooper pair at the sites 1 and N is drown in Fig. 3.12. The reason why the two states in the superposition acquire different phases is that the electron pairs in two states undergo evolutions in opposite directions of each other–this is, in fact, the well-known Aharonov–Bohm phase [74]. So, if in the first state the pair takes one path (i.e. the electron pair from site N moves to site 1), in the second state it takes the reverse of the same path (i.e. the electron pair from site 1

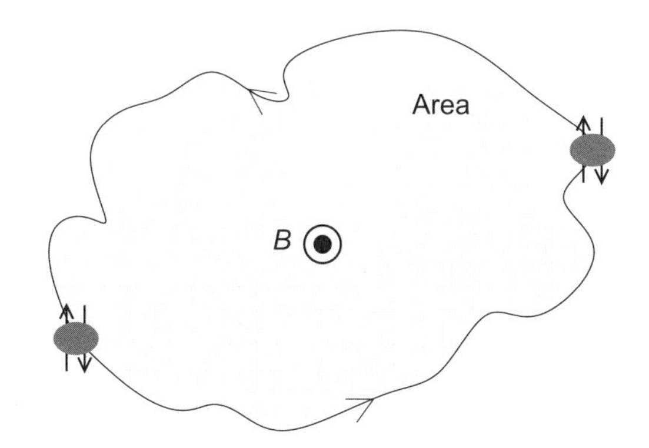

Fig. 3.12. Magnetic field and area.

moves to site N). The off-diagonal element in the $|0\rangle, |1\rangle$ basis of the two-site density matrix of sites 1 and N undergoes the following transformation $c^\dagger \to e^{i\Phi}c^\dagger$. However, the overall state must be totally symmetric, and therefore

$$e^{i\Phi} = 1. \tag{3.121}$$

From this we can conclude that

$$\Phi = \frac{2e}{\hbar c}\int Adl = \frac{2e}{\hbar c}\int\int BdS = 2n\pi, \tag{3.122}$$

where n is arbitrary integer. But, in the 2D (3D) space, the electron pair can take any trajectory, and the only possible choice that satisfies the above is $B = 0$ and $n = 0$ for an arbitrary A. Therefore, in a connected region of a material exhibiting bipartite entanglement there is no magnetic field present. This is the Meissner effect. We note that it is well-known that the magnetic field does penetrate the superconductor to a very small degree (falling off exponentially with the distance from the surface). Under the picture which has been explained, the effect is the result from the correlation between the Copper pair in the different sites having long-range order. It agrees with the picture given in the London equation as such

$$\nabla \times j + \frac{\rho_s e^2}{m_e}B = 0 \quad \text{for superconductor}, \tag{3.123}$$

which is with the electron density of superconductor ρ_s. On the contrary if it is normal conductor, they have different electron density does not necessary give the trivial magnetic field inside of the material which means $\nabla \times j + \frac{\rho e^2}{m_e}B = \text{Constant}$.

3.8. Boltzmann Equation

So far we have discussed the case that electron density is a constant over the time and the total number of particles are conserved at each different time. However, in many cases, large number of particles, e.g. atoms in a gas and electrons, interact each other and they are to be described in terms of the rate of change of their state in time. The motion of particles is influenced by their collisions with other particles although their net migration is mainly influenced by applied forces and by density gradients. The description of such behavior must be treated by the non-equilibrium problem and time dependent perturbation theory.

3.8.1. Prerequisite — Fermi's golden rule

When the Hamiltonian of an interacting system is not a constant with respect to time, the transition probability of the system must be averaged over time t. Consider the time-dependent Hamiltonian $H = H_0 + H_{\rm int}$ with the time-independent part of the Hamiltonian H_0 together with the time-dependent part of the Hamiltonian $H_{\rm int} = H_{\rm int}(t)$. In the interaction picture of the Hamiltonian, the transformed Hamiltonian $H_I = {\rm e}^{iH_0t/\hbar}H_{\rm int}{\rm e}^{-iH_0t/\hbar}$ satisfies the Schrödinger equation $H_I|\psi(t)\rangle = i\hbar\frac{\partial}{\partial t}|\psi(t)\rangle$.[p] The solution of the Schrödinger equation provides the state of the systems at a given time as

$$|\psi(t)\rangle = U(t)|\psi(0)\rangle \tag{3.124}$$

when the propagator $U(t)$ evolutes the state from an initial state $|\psi(0)\rangle$ to a final state $|\psi(t_f = T)\rangle$ as a solution of the operator differential equation $i\hbar\frac{\partial U(t)}{\partial t} = H_I U(t)$. Given that $U(0) = \mathbb{1}$, the explicit form of the propagator in the first-order expansion is $U(t) = \mathbb{1} + \frac{1}{i\hbar}\int_0^t H_I(t')U(t')dt'$ and the iterative substitution of the propagator generates the solutions in an arbitrary order as

$$\begin{aligned} U(t) &= \mathbb{1} + \frac{1}{i\hbar}\int_0^t H_I(t_1)dt_1 + \left(\frac{1}{i\hbar}\right)^2\int_0^t dt_1\int_0^{t_1} dt_2 H_I(t_1)H_I(t_2) + \cdots \\ &= \sum_{n=0}^{\infty}\left(\frac{1}{i\hbar}\right)^n\int_0^t dt_1\int_0^{t_1} dt_2\cdots\int_0^{t_{n-1}} dt_n\;\; H_I(t_1)H_I(t_2)\cdots H_I(t_n), \end{aligned}$$

which provides the exact formula of the state at the arbitrary time t. One should be careful about the order of the Hamiltonian $H_I(t)$ with different t since they are operators and they are not commute each other in general. Defining time ordering operator $\mathcal{T}\{\hat{A}(t_1)\hat{B}(t_2)\} \equiv \theta(t_1 - t_2)\hat{A}(t_1)\hat{B}(t_2) + \theta(t_2 - t_1)\hat{B}(t_2)\hat{A}(t_1)$ with the step function $\theta(t) = 1$ (for $t > 0$), $1/2$ (for $t = 0$), 0 (for $t < 0$) the propagator can be written in a simpler form,

$$U(t) = \mathcal{T}({\rm e}^{\frac{1}{i\hbar}\int_0^t dt' H_I(t')}), \tag{3.125}$$

where $\mathcal{T}$ denotes the defined *time ordering operator*. In the same way, the time evolution from the initial time t_i to the final time t_f can be found by replacing the propagator to two-time correlator as $U(t_f, t_i) =$

[p]The Schrödinger equation is given in terms of interaction picture. The eigenstate can be obtained from the solutions of the original Hamiltonian as $|\psi(t)\rangle = {\rm e}^{iH_0t/\hbar}|\psi(t)\rangle_S$ where $|\psi(t)\rangle_S$ is the eigenstate of the original Schrödinger equation $H|\psi(t)\rangle_S = i\hbar\frac{\partial}{\partial t}|\psi(t)\rangle_S$.

$\mathcal{T}(e^{\frac{1}{i\hbar}\int_{t_i}^{t_f} dt' H_I(t')})$. The solution of time-dependent Schrödinger equation can also be used to obtain the partition function of a static system at a finite temperature as it is discussed in Appendix A.7.

According to the time-dependent perturbation theory, the transition probability between the states can be evaluated. The higher order terms are negligible when the strength of the time-dependent interaction Hamiltonian is relatively small. When $H_{\text{int}}(t) = \lambda \hat{V} e^{-i\omega t}$ with $\lambda \ll 1$, the transition probability between an initial state $|\psi_i\rangle$ and a final state $|\psi_f\rangle$ at $t = T$ up to the first order becomes

$$p_{if} = |\langle \psi_f | U(T) | \psi_i \rangle|^2 = \left| \left\langle \psi_f \left| 1 + \frac{1}{i\hbar} \int_0^T e^{iH_0 t} H_{int} e^{-iH_0 t} dt \right| \psi_i \right\rangle \right|^2$$
$$= V_{if}^2 \frac{4 \sin^2 \left[\frac{(\omega - \omega_{if})}{2} T \right]}{(\omega - \omega_{if})^2}, \tag{3.126}$$

where $\omega_{if} = (E_f - E_i)/\hbar$ is the resonance frequency between the initial state and the final state. It is interesting to note that the transition probability is a function of difference between the resonance frequency and the oscillation frequency of the Hamiltonian ω, $\Delta \equiv (\omega - \omega_{if})/2$ as $p_{if} = V_{if}^2 \sin^2(\Delta T)/\Delta^2$. The probability function is sharply peaked at $\Delta = 0$ and becomes delta function in the limit $T \to \infty$.[q]

Suppose that the frequency density of perturbed Hamiltonian $H_{\text{int}}(t)$ is in the regime of broad driving or broadband with the density $g(\omega)$. In that case, the transition probability is needed to be averaged over the frequency density. Because the range of the frequency is vary narrow, the frequency density $g(\omega)$ can be considered as a constant near the peak value of p_{if}. Then the averaged probability density is given by

$$P = \int_{\Delta\omega} g(\omega) p_{if} d\omega \simeq g(\omega_{if}) V_{if}^2 T. \tag{3.127}$$

The result is significant in that the probability has a linear dependence in time. The transition rate is

$$\text{Transition rate} = \frac{dP}{dT} = g(\omega_{if}) V_{if}^2, \tag{3.128}$$

[q]The behavior of the sinusoidal function is $\lim_{T\to\infty} \frac{1}{\pi} \frac{\sin^2 T\Delta}{T\Delta^2} = \delta(\Delta)$ where δ is Dirac delta function. The function is normalized as $\int_{-\infty}^{\infty} \sin^2 x / x^2 = \pi$.

where $V_{if} = \langle\psi_f|\hat{V}|\psi_i\rangle$. It is the result which is known as *Fermi's Golden Rule*. This equation states that in order to calculate the transition rate all we have to do is multiply the square modulus of the perturbation matrix element between the two states, by the density of the states at the transition frequency. On the line it is also worthy to note that the transition probability is symmetric such as $P(i \to f) = P(f \to i)$.

3.8.2. *Transport-Boltzmann equation*

Originally, Boltzmann transport equation is developed in classical molecular gas theory [131] to find the transport coefficients such as the viscosity, the heat conductivity and so on. Later, the equation has been developed to investigate various non-equilibrium systems including electron–photon, electron–electron scattering problems. The first question in the problem is to consider the dynamics of a large number of particles (e.g. electrons or atoms in a gas) which interact each other and are scattered after the collisions. The aim is to describe the rate of their state changes in time and this is clearly the problem of the non-equilibrium systems.

The Boltzmann transport equation provides formula how the densities of large number of particles are changed as they are drifted in the gas or liquid. If the collision of the particles is taken into account, the equation of the systems becomes [123]

$$\left(\frac{\partial}{\partial t} + \vec{v}\cdot\nabla_{\vec{r}} + \frac{\vec{F}}{m}\nabla_{\vec{v}}\right) f(\vec{r},\vec{v},t) = \left(\frac{\partial f}{\partial t}\right)_{\text{coll}}, \tag{3.129}$$

where $\vec{F}$ is the force applied to the particles and $\vec{v}$ is the velocity of the particle. The essential step then is to find an explicit expression for $\left(\frac{\partial f}{\partial t}\right)_{\text{coll}}$. Boltzmann solved this problem under the simplifying assumptions that, (i) only binary collisions need to be considered (dilute gas); (ii) the influence of container walls may be neglected; (iii) the influence of the external force $\vec{F}$ (if any) on the rate of collisions is negligible; (iv) velocity and position of a molecule are uncorrelated (assumption of molecular chaos). The effect of the binary collisions is expressed in terms of a differential scattering cross section $\sigma(\Omega)$ which describes the probability density for a certain changes of velocities, $\{\vec{v}_1, \vec{v}_2\} \to \{\vec{v}_1', \vec{v}_2'\}$. The function $\sigma(\Omega)$ depends on the intermolecular potential and may be either calculated or measured.

Under all these assumptions and by a linear expansion of the left-hand side of the equation with respect to time, the Boltzmann equation takes on the following form:

$$\left(\frac{\partial}{\partial t} + \vec{v}_1 \cdot \nabla_{\vec{r}} + \frac{\vec{F}}{m}\nabla_{\vec{v}_1}\right) f(\vec{r}, \vec{v}_1, t) = \int d\Omega d\vec{v}_2 \frac{d\sigma}{d\Omega}|\vec{v}_1 - \vec{v}_2|(f_1' f_2' - f_1 f_2), \tag{3.130}$$

where $f_1 \equiv f(\vec{r}, \vec{v}_1; t)$, $f_1' \equiv f(\vec{r}, \vec{v}_1'; t)$, etc. Due to the energy conservation, $\vec{v}_1'$ is the function of $\vec{v}_1$ such that the left-hand side is only the function of $\vec{v}_1$. This integro-differential equation describes, under the given assumptions, the spatio-temporal behavior of a dilute gas. Given some initial density $f(\vec{r}, \vec{v}; t = 0)$ in μ space[r] the solution function $f(\vec{r}, \vec{v}; t)$ tells us how this density changes over time. Since f has up to six arguments, it is difficult to visualize; but there are certain moments of f which represent measurable averages such as the local particle density in 3D space, whose temporal change can thus be computed. In general the solution is not easy to compute but we dealt with the problem using rate equation in the next.

It is possible to derive the Boltzmann transport equation above from the classical theory. Let us consider the molecules in the infinitesimal volume $d\vec{r}d\vec{v}$, especially for the first molecule it is denoted as $d\vec{r}_1\ d\vec{v}_1$. During the time between t and $t + \delta t$, the molecules in this volume element collide each other such that some molecules are kicked in and others are kicked out of the volume element. As the assumption is made, we only consider the binary collision between two momentums $\vec{v_1}$ and $\vec{v_2}$. In this case, the number of transitions $12 \to 1'2'$ in a volume element $d\vec{r}$ is given by $dN_{12}dP_{12\to 1'2'}\delta t$, where dP is the transition rate in Eq. (3.128) and dN_{12} is the number of particles before the collision. With the two-particle correlation $F(\vec{r}, \vec{p}_1, \vec{p}_2, t)$, dN_{12} is described as

$$dN_{12} = F(\vec{r}, \vec{p}_1, \vec{p}_2, t)d\vec{r}d\vec{p}_1 d\vec{p}_2. \tag{3.131}$$

In terms of the volume element $d\vec{r}_1 d\vec{v}_1$ for the particle 1, the total rate of binary collisions in which molecules kicked out of the volume element

[r]Vector space of six independent parameters describing a particle. It is followed by three for the position and three for the velocity.

$d\vec{r}_1 d\vec{v}_1$ is given by

$$R_{\text{out}} = \int d\vec{p}_2 d\vec{p}_1^{'} d\vec{p}_2^{'} \, \delta^4(\mathbf{P}_f - \mathbf{P}_i)|T_{fi}|^2 F(\vec{r}, \vec{p}_1, \vec{p}_2, t) \tag{3.132}$$

and for the molecules kicked in the volume element is

$$R_{\text{in}} = \int d\vec{p}_2 d\vec{p}'_1 d\vec{p}'_2 \, \delta^4(\mathbf{P}_i - \mathbf{P}_f)|T_{if}|^2 F(\vec{r}, \vec{p}_1, \vec{p}_2, t), \tag{3.133}$$

where the $\delta^4(\mathbf{P}_f - \mathbf{P}_i)|T_{fi}|^2$ (and $f \to i$) can be inferred from Eq. (3.127) and $\mathbf{P}$ is denoted by the total momentum of two molecules. Due to the rotational, reflection, and inversion symmetry, the elements of transition matrix T are the same, $|T_{if}| = |T_{fi}|$, and two δ functions in R_{out} and R_{in} the same. Therefore,

$$\begin{aligned} \left(\frac{\partial f}{\partial t}\right)_{\text{coll}} &\equiv R_{\text{out}} - R_{\text{in}} \\ &= \int d\vec{p}_2 d\vec{p}'_1 d\vec{p}'_2 \delta^4(P_f - P_i)|T_{fi}|^2 (F_{1'2'} - F_{12}), \end{aligned} \tag{3.134}$$

where $F_{12} = F(\vec{r}, \vec{p}_1, \vec{p}_2, t)$. If we embrace the correlation between the momenta of two molecules, then the two-particle correlation function is simplified just as the product of the probability of two molecules

$$F(\vec{r}, \vec{p}_1, \vec{p}_2, t) = f(\vec{r}, \vec{p}_1, t) f(\vec{r}\vec{p}_2, t), \tag{3.135}$$

Finally, we obtain the Boltzmann transport equation

$$\left(\frac{\partial f}{\partial t}\right)_{\text{coll}} = \int d\vec{p}_2 d\vec{p}'_1 d\vec{p}'_2 \, \delta^4(P_f - P_i)|T_{fi}|^2 (f_{1'} f_{2'} - f_1 f_2). \tag{3.136}$$

If we remind the process of the scattering, then one can obtain the $dP_{12 \to 1'2'} = I d\sigma$ with the incident flux I and it is enough to derive Eq. (3.130) [123].

In the quantum treatment of the question, we retain two assumptions to be used for the simplification of the problem. At first, we consider only the binary elastic collisions which is involved with a static potential $V(x - x')$ during the collision. Secondly only the first-order perturbation is important so that the transition probability follows Fermi's golden rule. Under the circumstance, the number of particles $N(k) = f(\vec{r}, \vec{v}; t) d\mu$ having specific momentum k is subject to the interactions between the particles having different momentum. The fluctuation of the particle number over the time

$\frac{\partial N(k)}{\partial t}$ is linearly proportional to the number of scattered particles and the occupation number of the scattering state. It will be described by the rate equation:

$$\frac{\text{prob}}{\text{time}} \propto \frac{2\pi}{\hbar}|\nu(k)|^2 N(k+q)N(k'-q) \times [1 \pm N(k)][1 \pm N(k')] \\ \times \delta\left[\frac{\hbar^2}{2m}(|k+q|^2 + |k'-q|^2 - k^2 - k'^2)\right], \tag{3.137}$$

where $|\nu(k)|$ is strength of interaction, $+/-$ is for boson/fermion and the last term means energy conservation. The probability of the particle fluctuation is also influenced by the scattering in the opposite direction and the contribution needs to be subtracted from the rate equation (Fig. 3.13). Taken them all into account, quantum Boltzmann equation becomes

$$\frac{\partial N(k)}{\partial t} = \sum_{k'}\sum_{q}\frac{2\pi}{\hbar}|\nu(q)|^2 \cdot \delta(\text{energy}) \\ \times \{N(k+q)N(k'-q)[1 \pm N(k)][1 \pm N(k')] \\ - N(k)N(k')[1 \pm N(k+q)][1 \pm N(k'-q)]\}. \tag{3.138}$$

The delta function $\delta(\text{energy}) = \delta(|k+q|^2 + |k'-q|^2 - k^2 - k'^2)$ imposes the condition of the energy conservation such that $q = k' - k$. It confirms that the rate equation approaches to the equilibrium solution of the particle density $\frac{\partial N(k)}{\partial t} = 0$ in the scattering process. The number of fluctuation around the equilibrium at temperature T is found from the rate equation

$$\frac{\partial N(k)}{\partial \varepsilon}\frac{\partial \varepsilon}{\partial t} \propto N(k)[1 \pm N(k)] \tag{3.139}$$

as such the number of distributions becomes

$$N(k) = \frac{1}{Ae^{\frac{\hbar^2 k^2}{2mk_B T}} \mp 1}, \tag{3.140}$$

where k_B is the Boltzmann constant.

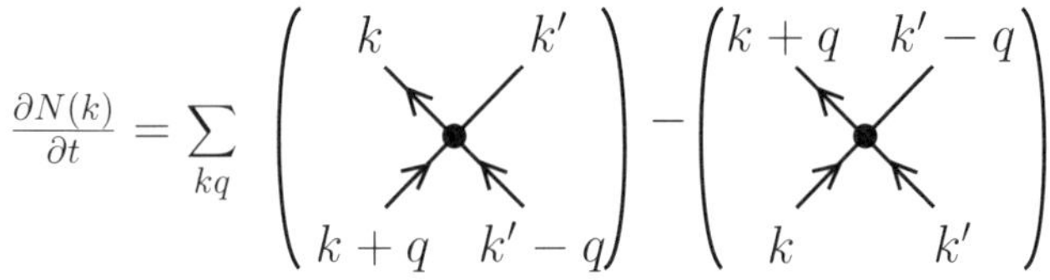

$\frac{\partial N(k)}{\partial t} = \sum_{kq}$ (k k' / $k+q$ $k'-q$) $-$ ($k+q$ $k'-q$ / k k')

Fig. 3.13. Rate equation with scattering.

The rate equation from the Fermi's golden rule bears strong implication for the derivation of the second law of thermodynamics. The second law of thermodynamics says that entropy of an isolated system never decreases. In statistical mechanics, an isolated system can be modeled as a Markov chain. Suppose that a system can take a state $1, 2, \ldots$ each one occupied with the initial probabilities $P_1, P_2, \ldots$ respectively. The evolution of this physical system is now represented by transition probabilities of the form $W_{i,f}$ that the system in the initial state i will jump to the final state f. The rate changes of the particular probability is given by

$$\frac{dP_f}{dt} = \sum_i \{W_{i,f}P_i - W_{f,i}P_f\}. \tag{3.141}$$

The relation is known as Fermi's matter equation. Since the underlying physics for the jumping process is reversible, it turns out that the transition probability is symmetric $W_{i,f} = W_{f,i}$. In fact, it is the direct consequence of Fermi's golden rule. The master equation on the other hand leads to irreversible processes. From the definition of entropy $S = -k\sum_i P_i \ln P_i$, the time derivative of the entropy reads

$$\frac{dS}{dt} = -k\sum_i \left\{\frac{dP_i}{dt} + \frac{dP_i}{dt}\ln P_i\right\} = -k\sum_i \frac{dP_i}{dt}\ln P_i \tag{3.142}$$

since $\sum_i P_i = 1$ and thus $\frac{d}{dt}(\sum_i P_i) = 0$. Then, using the master equation with the jump rate symmetry, one would have that

$$\frac{dS}{dt} = -\frac{k}{2}\sum_{i,f} W_{i,f}(P_f - P_i)(\ln P_f - \ln P_i), \tag{3.143}$$

which is positive function for any values of P_i and P_f. Hence this proves that

$$\frac{dS}{dt} \geq 0 \tag{3.144}$$

and the inequality shows that the entropy is increasing all the time.

3.9. Problems

1. **Two-coupled harmonic oscillators:** The Hamiltonian for two-coupled quantum simple harmonic oscillators is given by $H = \Omega(a^\dagger a + b^\dagger b) + V(a^\dagger b + b^\dagger a)$. Diagonalize this Hamiltonian to find all the eigenvalues and eigenstates. If the system starts in the state $|1, 0\rangle$ (where

the first slot corresponds to the mode a and the second to the mode b, what is its state at some later time t?

2. **Hartree–Fock:** Derive the exchange integral for jellium in the Hartree–Fock mean field theory approximation.
3. **Group theory and fermions:** Prove that permutations of three elements (call them 1,2,3) form a group. Is the resulting group Abelian (i.e. commutative)? Write down all possible states of three electrons (both space and spin components of the wave function) assuming that they are completely indistinguishable and that the spatial part is antisymmetrized.
4. **Pauli paramagnets:** Take a spin in a magnetic field in the z direction. Calculate its (average) internal energy at temperature T and use this to infer the resulting heat capacity. Now assume that the two energy levels are broadened equally to some value Δ. Usually some kind of external noise is responsible for the broadening, but we are here not concerned about the details. Show that the heat capacity is now linearly proportional to temperature provided that $kT \ll \Delta$. This result is very useful in the theory of glasses.
5. **Ising ferromagnets:** Imagine two-coupled electron spins through the Ising Hamiltonian $J\sigma_{1z}\sigma_{2z}$, where the subscript labels the spin and the superscript the direction of interaction. What are the eigenstates and eigenenergies of the system? Which state will the system occupy at zero temperature? Now assume that an external field of strength B is applied in the x direction. Write down the resulting Hamiltonian and obtain new eigenvectors and eigenvalues. What state is now occupied at zero temperature as a function of B and J?
6. **Superconductivity and long-range order:** We can think of the (single Cooper pair) superconducting ground state as a coherent superposition of Cooper pair states $|\psi\rangle = \int \psi_\uparrow(x)\psi_\downarrow(x)|0\rangle$, where $|0\rangle$ is the empty state (containing no electrons). Prove that this state exhibits the phenomenon of long-range order, namely that $\langle\psi|\psi^\dagger(x)\psi^\dagger(x)\psi(y)\psi(y)|\psi\rangle$ is not equal to zero for any x and y (it is usually state "as x and y become very far apart"). What would the state with N Cooper pairs look like? Would it also possess long-range order and why?
7. **Transport equation and entropy:** Starting with the Pauli Master equation, show that the entropy of thus described system can never decreases in time. This is the best "derivation" we have of the second law of thermodynamics.

Chapter 4

Contemporary Topics in Quantum Many-Body Physics

Quantum property in many-body systems is manifested when the individual constituents are correlated and entangled. The problem of measuring entanglement is a vast and lively field of research in its own and numerous different methods have been proposed for its quantification. In this chapter, we provide the role of many-body entanglement for exotic quantum phenomena and their identification. It is relatively recent to investigate the entanglement structure in many-body system and the area is still under serious investigation.

Here, we are trying to avoid an exhaustive review of the entanglement studies. Rather, we introduce those measures that are largely being used to quantify entanglement in many-body systems. Comprehensive overviews of entanglement measures can be found in [56, 135–141]. In this context, we also outline a method of detecting entanglement, based on entanglement witnesses and their effect on the macroscopic quantity of the system. We start from the extensive introduction in the following section.

4.1. When Do Superfluidity and Long-Range Order Imply Entanglement?

Phase transitions play a very important role in physics because they challenge us to describe these inherently macroscopic phenomena in terms of the underlying, microscopic, mechanisms [18]. We have many different types of phase transitions, ranging from a solid to liquid to gas transition of large numbers of atoms, to the quantum mechanical electron (Cooper) pairing signifying the onset of superconductivity. All different phase transitions, however, have one crucial aspect in common. Simply put, by varying external conditions, a certain same lump of matter (or light) can fundamentally change its macroscopic behavior. Ehrenfest was the first to classify phase transitions according to which macroscopic property in particular suffers this abrupt change. If the first derivatives of the free energy (such as the internal energy, or magnetization) are the observables in question, then the phase transition is of the first order; if the second derivatives become discontinuous (e.g. magnetic susceptibility or heat capacity), then the phase transition is of the second order, and so on. In spite of this advancement, a coherent account of phase transitions was still amiss until the application of quantum mechanics to many-body physics.

The first theory of superfluid critical phenomena was developed by Landau [20] and also goes under the name of mean-field theory. It suggests that any phase transition can be viewed as an order to disorder transition. To quantify it, we therefore need to identify a function (defined at all points in space), called an order parameter, whose non-zero value identifies that we are below the critical point (e.g. temperature) where order exists. Above the critical temperature, the order parameter disappears. The existence of a function that stretches across the whole physical system in the ordered phase immediately suggests that different parts of the system, even if far apart, are in some sense correlated through this order parameter function. This gives rise to the notion of long-range order (LRO), introduced by Penrose and Onsager [21] in order to understand Bose condensation. The condition for the existence of LRO is given by

$$\lim_{|x-x'|\to\infty} \langle \Psi^\dagger(x)\Psi(x')\rangle \Longrightarrow \text{const} > 0 \tag{4.1}$$

and for Penrose and Onsager is identical to the system being in a Bose condensate. The constant to which the two-point correlations tend in the large size of the system limit is, in fact, the order parameter discussed before. For us here, the interesting question will be if this long-range order implies the existence of entanglement and if so, what kind of entanglement might that be. Before that, let us first finalize our brief exposition of phase transitions.

By the time low temperature superconductivity was properly described by Bardeen, Cooper and Schrieffer [22], it was clear that the LRO does not quite apply to it immediately. Electrons in a superconductor are not correlated across such large distances that they cover the whole of the superconducting sample. Yang was the first to realize that we need a new notion of order which he termed the "off-diagonal long-range order" (ODLRO) [23]. If instead of $\Psi(x)$ representing a single electron at position x, this operator represented a Cooper pair at the same x, $\Psi(x) = \psi_\uparrow(x)\psi_\downarrow(x)$, then the Penrose Onsager condition for LRO would still be satisfied by this new operator. The proper order parameter is therefore the wave function for a Cooper pair — a pair of electrons in a spin singlet state. We can then think of a superconductor as a condensate of Cooper pairs, though electrons are still fermions and behave as such. (In BCS theory of superconductivity, we can likewise think of the energy gap between the ground state with Cooper pairs and the excited states where they start to break up as a good order parameter. The gap disappears exactly above the critical temperature, when the superconductor transforms into an ordinary conductor. Both of these formulations are equivalent and lead to the same calculation of the critical temperature for superconductivity.)

The next big surprise in understanding phase transition came in 1970s when it was discovered that we can actually have a phase transition without the existence of any long-range order. This discovery was, interestingly, predated by a number of different proofs of the fact that continuous phase transitions cannot exist in 1 and 2 spatial dimensions [24–26]. The reason for this is very simple and it lies in the fact that any long-range order is destroyed by thermal fluctuations in low dimensions. However, what can happen is that a kind of short-range order emerges in two dimensions (we will refer to this a quasi-long-range order) out of a disordered phase. This means that below some temperature, the correlations exhibit a polynomial drop with the distance like $\lim_{|x-x'|\to\infty}\langle\Psi^\dagger(x)\Psi(x')\rangle \longrightarrow 1/|x - x'|^p$, where p is some power. Above this temperature the drop is typically exponential. The mechanism for this transition was explained by Berezinskii, Kosterlitz and Thouless [27] (BKT) and has been experimentally verified a number of times since. In summary, therefore, we can have long-range order, off-diagonal long-range order as well as quasi-long-range order present in matter and all linked to phase transitions.

The whole story about order–disorder transitions can and has been applied to quantum phase transitions as well [28]. These transitions occur at zero temperature (in practice, at low temperatures) and are driven by changes of some parameter other than temperature, such as the external magnetic field in a spin chain or a doping parameter in a high temperature superconductor. Quantum phases can also exhibit long-range order and quasi-long-range order. The point here, of course, is that any correlations are now likely to imply some type of entanglement, since we have an overall pure state [29]. In the same way that different correlations in three dimensions lead to different crystals, we might expect that different types of entangled state will lead to different phenomena in quantum critical regions.

4.1.1. *Independence of entanglement and quantum order*

We address here the issue of how quantum criticality and entanglement are related. We find that the relationship is not entirely straightforward, but some concrete conclusions can nevertheless be drawn.

Before we start our general discussion, we note that entanglement and quantum ODL are seemingly not strictly related to one another: each can exist without the presence of the other. For example, the product of coherent states of equal amplitude but at different positions x, $\prod_x |\alpha(x)\rangle$, is spatially a disentangled state (by construction), but it exhibits ODLRO (equal to $|\alpha|^2$). On the other hand, the state of the GHZ type $|000..0\rangle + |111..1\rangle$ is clearly entangled, but LRO does not exist. The latter can be seen by the fact that the operator combination $\psi^\dagger\psi$ vanishes for states with the same number of excitations such as $|00\rangle$ and $|11\rangle$. (Here, it is more appropriate to introduce Pauli raising and lowering operators, but we leave this for later.) Therefore, any kind of criticality that is indicated by LRO can be very different from the criticality signified by the vanishing of entanglement. Similar statements can be made about entanglement and quasi-long-range order. Entanglement is, in fact, just a more complex form of quantum order, and, just like correlations, it can exist between any collection of spins, either they are close to each other or on different sides of the investigated system (for a recent review of entanglement in many-body systems, see [30]).

If, as above, we can show that entanglement and LRO are independent, why has there been so much work on entanglement and phase transitions? The simple answer is that states do not encode all the information about the actual physics of the system. The Hamiltonian plays an equally important role. It can easily happen that the states we discussed above (coherent and GHZ) are never the eigenstates of the relevant Hamiltonian. This is exactly what will happen in the situations analyzed below. We will show that entanglement is, in spite of the above arguments to the contrary, intimately connected to quantum order and critical phenomena.

Our approach here will entirely be based on physically observable effects. The notions of different order parameters are useful precisely because they are ultimately linked to some underlying physics. Long-range order in superconductors, for example, has been shown to imply the Meissner effect as well as the notion of flux quantization [31] (in superfluids, it leads to irrotational flow and quantized vortices [32]). It is in this sense that it would be desirable to view entanglement: can we say that superconductivity (-fluidity), or some other critical phenomena, are in any way dependent on entanglement? In the following, we analyze this question in detail and show that the answer is affirmative. Our example will be very simple, but it contains all the necessary elements to draw some general conclusions. We will analyze both classical (i.e. $T > 0$) and quantum ($T = 0$) criticality and show that recent experiments with cold atoms already confirm the existence of field theoretic, particle number, entanglement.

4.1.2. *Experimental considerations for spin entanglement*

How would we confirm the existence of spatial entanglement experimentally? To be able to do so, we need to extract the value of the nearest neighbor two-point correlation function $\langle \sigma_i^+ \sigma_{j+1}^- \rangle$. One way of estimating this is by interfering independent fluctuating condensates [47].

Let us sketch the recent remarkable experiment of Hadzibabic *et al.* [48] and investigate if their experiments reveal any entanglement in two-dimensional (2D) Bose gases by interfering independent condensates. In their experiments, the value of the chemical potential is $\mu/h = 10KHz$, while the (average) separation between the atoms will be taken to be less than or equal to the healing length $a = 0.2\,\mu$m (as is true for a dilute Bose gas analyzed here, see the Appendix). The temperatures reached during the experiment were 100–200 nK. They used Rubidium atoms whose mass is 87 atomic units. Putting these numbers together, we obtain that $\mu^2 + (kT)^2 < J^2$, which as we showed was our criterion for the existence of entanglement. Therefore, Bose condensates exhibit field theoretic (continuous variable) entanglement (previously, we have only had evidence of spin-based entanglement in many-body systems [41, 49]).

Furthermore, we see that the onset of entanglement is very closely related to the BKT transition. The BKT transition was measured to occur at temperatures for which the thermal de Broglie wavelength is $\lambda_T = 0.3\,\mu$m. The healing length, which we can think of as the effective lattice spacing, is $a = 0.2\,\mu$m. Entanglement occurs at $a < \lambda$ which agrees with the BKT transition. This can also be obtained from the aforementioned field theoretic criterion [46] that $kT < J = \hbar^2/(2ma)$, where a is the healing length. What is more, we can even estimate the amount of entanglement, as quantified by the single spin reduced entropy. Since $\epsilon_\pm = 1 \pm (1 - 2/\pi \arccos(\mu/J))$, the (single spin) entanglement is $E \approx 0.33$. The fact that entanglement seems to occur at the point where the de Broglie wavelength becomes larger than the healing length, which itself physically represents the core size of a vortex, would suggest that entanglement may be linked to the vortex–anti vortex pairing that characterizes the BKT superfluid phase. This point requires a further more in-depth study.

4.1.3. *A few further remarks*

As we have seen, the traditional indicators of order in matter are, superficially speaking, not necessarily strictly related to the existence of entanglement between the underlying constituents. However, when the whole situation is properly analyzed using the information given by a system's Hamiltonian, we see that some rigorous relationships naturally emerge. We have found conditions for which superfluidity implies entanglement and then demonstrated that some recent experiments have already been sufficiently detailed to witness entanglement in Bose gases.

Furthermore, based on very general assumptions, it can be shown that no long-range order is possible in one and two dimensions, while it is clearly possible to have entanglement under the same circumstances. It is therefore tempting to speculate that entanglement can be related to low-dimensional quantum criticality other than the other known order parameter. The present discussion substantiates this view with an entanglement analysis of recent cold atom experiments investigating the BKT transitions. Using the entanglement for practical purposes would be the next desirable item to investigate in greater detail.

4.2. High Temperature Macroscopic Entanglement

Entanglement is currently one of the most researched phenomena in physics. Often shrouded in mystery, its basic premise is quite simple — entanglement is a correlation between distant particles that exists outside of any description offered by classical physics. Predictions from the theory of entanglement have confounded some of the greatest minds in science. Einstein famously dubbed it spukhafte Fernwirkungen: "spooky action at a distance". As we look deeper into the fabric of nature this "spooky" connection between particles is appearing everywhere, and its consequences are affecting the very (macroscopic) world that we experience. At an implementational level, using entanglement researchers have succeeded in teleporting information between two parties, designing cryptographic systems that cannot be broken and speeding up computations that would classically take a much longer time to execute [52]. Even though these applications have generated significant interest, we have only scratched the "tip of the iceberg" in terms of what entanglement is, and indeed what we can do with it.

While entanglement is experimentally pretty much beyond dispute in microscopic systems — such as two photons or two atoms — many people find it difficult to accept that this phenomenon can exist and even have effects macroscopically. Based on our everyday intuition we would, for example, find it very hard to believe that two cats or two human beings can be quantum entangled. Yet quantum physics does not tell us that there is any limitation to the existence of entanglement. It can, in principle and as far as we understand, be present in systems of any size and under many different external conditions.

The usual argument against seeing macroscopic entanglement is that large systems have a large number of degrees of freedom interacting with the rest of the universe and it is this interaction that is responsible for destroying entanglement. If we can exactly tell the state that a system is in, then this system cannot be entangled to any other system. In everyday life, objects exist at room (or comparable) temperatures so their overall state is quantum mechanically described by a very mixed state (this mixing due to temperature is, of course, also due to the interaction with a large "hot" environment). Mixing states that are entangled, in general, reduces entanglement and ultimately all entanglement vanishes if the temperature is high enough. The question then is how high is the highest temperature before we no longer see any entanglement? And how large can the body be so that entanglement is still present? Can we, for example, have macroscopic entanglement at the room temperature?

Entanglement has recently been shown to affect macroscopic properties of solids, such as its magnetic susceptibility and heat capacity, but at a very low (critical) temperature [53]. This extraordinary result demonstrates that entanglement can have a significant effect in the macroscopic world.

The basic reason for this dependence is simple. Magnetic susceptibility is proportional to the correlation between nuclear spins in the solid. As we said before, entanglement offers a higher degree of correlation than anything allowed by classical physics and the corresponding quantum susceptibility — which fully agrees with experimental results [53] — is higher than that predicted by using just classical correlations (for further theoretical support for this see the article in [34]). It is now very important to go beyond this low temperature regime and experimentally test entanglement at higher temperatures.

Thinking that high temperature entanglement is linked with (perhaps even responsible for) some other high temperature quantum phenomena, such as high temperature superconductivity, is tempting. After all, superconductivity is a manifestation of the existence of the off-diagonal long-range order (ODLRO) [23] which is a form of correlation that still persists in the thermodynamical (macroscopic) limit. However, it is not immediately obvious that this correlation contains any quantum entanglement. The main intention in this section is to show that it does. This correlation contains multipartite entanglement between all electron pairs in the superconductor. To calculate this we need to be able to quantify entanglement exactly and be able to discriminate entanglement from any form of classical correlation.

A great deal of effort has gone into theoretically understanding and quantifying entanglement [55]. There are a large number of different proposed measures; the different measures capture different aspects of entanglement. Here, we are interested in a measure that is based on the (asymptotic) distinguishability of entangled states from separable (disentangled) states known as the relative entropy of entanglement [56, 57]. The main advantage of this measure is that it is easily defined for any number of systems of any dimensionality, which is not the case for entanglement of formation or distillation [55]. It can be argued that a number of results in quantum information and computation follow from the relative entropy function [55].

There is, unfortunately, no closed form for the relative entropy of entanglement, but this measure can still be computed for a large class of relevant states such as the pure bipartite states, Werner states and many others [57]. Most recently, Wei *et al.* [58] have succeeded in obtaining a formula for the relative entropy of entanglement for any number of totally symmetric pure states of n qubits using a very simple and elegant argument (some partial results have been obtained previously in this direction using different methods by Plenio and Vedral [59], but only for three qubit symmetric states). These results are used and extended further with the idea of applying them to a specific model of a superconductor.

The purpose of this discussion is to investigate possible links between high temperature entanglement and high temperature superconductivity with the intention of showing that entanglement can persist at higher temperatures. We analyze a particular mechanism — the η-pairing of electrons due to Yang [60] — that was originally proposed to explain high temperature superconductivity. The main difference between this pairing mechanism and the usual Bardeen, Cooper and Schrieffer (BCS) electron pairing [22] for (low temperature) superconductivity is that, in the former, electrons that are positioned at the same site are paired, while in the latter, electrons forming Cooper pairs are separated by a certain finite average distance (the so-called coherence length, typically of the order of hundreds of nanometers). The physical reason behind electron pairing is also thought to be different in a high temperature superconductor, but we do not enter into discussing these details here (see e.g. [61]). We, however, look at the η model in a different way, using totally symmetric states, and this will make calculating entanglement easier.

Wei *et al.* [58] have recently made very important steps in calculating the relative entropy of entanglement for symmetric state. Their approach is extended to calculating the relative entropy of entanglement for mixed symmetric state arising from tracing over some qubits in pure states, and apply it to understanding various relations between entanglements of a subset of qubits and their relation to the total entanglement. It will be shown that although two-site entanglement disappears as the distance between sites diverges (a conclusion also reached by Zanardi and Wang in a different way [72]), the total entanglement still persists in the thermodynamical limit. Furthermore, it scales logarithmically with the number of qubits. Therefore, it is this total entanglement that should be compared with ODLRO and not the two-site entanglement. While the two-site entanglement vanishes thermodynamically, two-site classical correlations are still present and so is the entanglement between two clusters of qubits (two cluster entanglement in η states has also been analyzed by Fan [63]). It also will be shown that all aspects of my analysis can easily be generalized to higher than half spin systems. The aim of this work — which is really just a first step in exploring high temperature entanglement — will be extended to different models with states other than symmetric and this will allow us a much more complete understanding of entanglement and the role it plays in the macroscopic world.

4.2.1. *η-pairing in superconductivity*

The model that is described now consists of a number of lattice sites, each of which can be occupied by fermions having spin up or spin down internal states. Let us introduce fermion creation and annihilation operators, $c^\dagger_{i,s}$ and $c_{i,s}$ respectively, where the subscript i refers to the ith lattice site and s refers for the value of the spin, $\uparrow$ or $\downarrow$. Since fermions obey the Pauli exclusion principle, we can have at most two fermions attached to one and the same site. The c operators therefore satisfy the anticommutation relations:

$$\{c_{i,s}, c^\dagger_{j,t}\} = \delta_{ij}\delta_{s,t} \tag{4.2}$$

and c's and $c^\dagger$'s anticommute as usual. (Some general features of fermionic entanglement — arising mainly from the Pauli exclusion principle — have been analyzed in [64, 66, 72, 73].)

We only need to assume that our model has the interaction which favors formation of Cooper pairs of fermions of opposite spin at each site [60]. The actual Hamiltonian is not relevant for my present purposes. It suffices to say that Yang originally considered the Hubbard model for which the η states are took account.

Here, we use and extend eigenstates (but none of them is a ground state [60]). A generalization of the Hubbard model was presented in [67] and in a specific regime of this new model the η states do become lowest energy eigenstates (this is a fact that will become relevant when we talk about high temperature entanglement). Both these models have been used to simulate high-temperature superconductivity, since in high superconducting materials, the coherent length of each Cooper pair is on average much smaller than for a normal superconductor.

Suppose, now, that there are n sites and suppose, further, that we introduce an operator $\eta^\dagger$ that creates a coherent superposition of a Cooper pair in each of the lattice sites,

$$\eta^\dagger = \sum_{i=1}^{n} c^\dagger_{i,\uparrow} c^\dagger_{i,\downarrow}. \tag{4.3}$$

The $\eta^\dagger$ operator can be applied to the vacuum a number of times, each time creating a new coherent superposition. However, the number of applications, k, cannot exceed the number of sites, n, since we cannot have more than one pair per site due to the exclusion principle. Accordingly we now introduce the following basis:

$$|k, n-k\rangle := \frac{1}{\sqrt{\binom{n}{k}}}(\eta^\dagger)^k|0\rangle, \tag{4.4}$$

where the factor in front is just the necessary normalization. Here, the vacuum state $|0\rangle$ is annihilated by all c operators, $c_{i,s}|0\rangle = 0$. We note in passing that the originally defined η operators can also have phase factors dependent on the location of the site on the lattice. We can have a set of operators like

$$\eta_k = \sum_n e^{ikn} c^\dagger_{n,\uparrow} c^\dagger_{n,\downarrow}. \tag{4.5}$$

All the states generated with any η_k from the vacuum will be shown to have the same amount of entanglement so that the extra phases will be ignored in the rest of the paper (i.e. we will only consider the $k = 0$ states).

We can think of the η states in the following way. Suppose that $k = 2$. Then this means that we will be creating two η-pairs in total, but they cannot be created in the same lattice site. The state $|2, n-2\rangle$ is therefore a symmetric superposition of all combinations of creating two pairs at two different sites. Let us, for the moment, use the label 0 when the site is unoccupied and 1 when it is occupied. Then the state $|2, n-2\rangle$ is

$$|2, n-2\rangle = \frac{1}{\sqrt{\binom{n}{2}}}(|\underbrace{000}_{n-2}\cdots\underbrace{11}_{2}\rangle + \cdots|\underbrace{11}_{2}\cdots\underbrace{000}_{n-2}\rangle), \tag{4.6}$$

i.e. it is an equal superposition of states containing 2 states $|1\rangle$ and $n-2$ states $|0\rangle$. These states, due to their high degree of symmetry, are much easier to handle than general arbitrary superpositions and we can compute entanglement for them between any number of sites. Note that in this description each site effectively holds one quantum bit, whose 0 signifies that the site is empty and 1 signifies that the site is full.

The main characteristic of η states is the existence of the long-range off diagonal order (ODLRO), which implies its various superconducting features, such as the Meissner effect and flux quantization [31]. The ODLRO is defined by the off diagonal matrix elements of the two-site reduced density matrix being finite in the limit when the distance between the sites diverges. Namely,

$$\lim_{|i-j|\to\infty} \langle c^\dagger_{j,\uparrow} c^\dagger_{j,\downarrow} c_{i,\downarrow} c_{i,\uparrow}\rangle \longrightarrow \alpha \tag{4.7}$$

where α is a constant (independent of n). It can be shown that although the existence of off diagonal matrix elements does not guarantee the existence of entanglement between the two sites, it does guarantee the existence of multi-site entanglement between all the sites. Note that here, by "correlations" it means correlations between the number of electrons positioned at different sites i and j. Namely, we are looking at the probability of one site being occupied (empty) given that the other site is occupied (empty). This is different from spin–spin correlations, which would look at the occurrences of both electron spins being up or down, or one being up and the other being down [64].

4.2.2. *General description of symmetric states*

The states we analyze here will always be of the form

$$|\Psi(n,k)\rangle \equiv |k, n-k\rangle := \frac{1}{\sqrt{\binom{n}{k}}}(\hat{S}|\underbrace{000}_{k}\ldots\underbrace{11}_{n-k}\rangle), \tag{4.8}$$

where $\hat{S}$ is the total symmetrization operator. We consider mixtures of these states, which become relevant when we talk about systems at finite temperatures. Symmetric states arise, for example, in the Dicke model in which n atoms simultaneously interact with a single mode of the electromagnetic field [68]. They are, furthermore, very important as they happen to be eigenstates of many models in solid-state physics, and, in particular, they are eigenstates of the Hubbard and related models supporting the η pairing mechanism. The analysis presented here will be applicable to any of these systems and not just the η model. The η mechanism will be significant because of its potential to support high temperature entanglement.

Let us start to compute the entanglement between every pair of qubits (sites) in the above state $|\Psi(n,k)\rangle$. A simpler task would be first to tell if and when every pair of qubits in a totally symmetric state is entangled. For this, we need only compute the reduced two-qubit density matrix which can be written as to

$$\sigma_{12}(k) = a|00\rangle\langle 00| + b|11\rangle\langle 11| + 2c|\psi^+\rangle\langle\psi^+|, \tag{4.9}$$

where $|\psi^+\rangle = (|01\rangle + |10\rangle)/\sqrt{2}$ and

$$\begin{aligned} a &= \frac{\binom{n-2}{k-2}}{\binom{n}{k}} = \frac{k(k-1)}{n(n-1)}, \\ b &= \frac{\binom{n-2}{k}}{\binom{n}{k}} = \frac{(n-k)(n-k-1)}{n(n-1)}, \\ c &= \frac{\binom{n-2}{k-1}}{\binom{n}{k}} = \frac{k(n-k)}{n(n-1)}. \end{aligned} \tag{4.10}$$

We can easily check that $a + b + 2c = 1$ and so the state is normalized. This density matrix is the same no matter how far the two sites are from each other, since the state is symmetric, and must therefore be identical for all qubits. We can easily test the Peres–Horodecki (partial transposition) condition [69] for separability of this state. Two states are entangled if and only if they are inseparable which leads to states $\sigma_{12}(k)$ being entangled if and only if

$$a + b - \sqrt{(a-b)^2 + 4c^2} < 0, \tag{4.11}$$

which leads to

$$(k-1)(n-k-1) < k(n-k). \tag{4.12}$$

This equation is satisfied for all $n \geq 2$ (two qubits or more) and $1 \leq k \leq n-1$. So, apart from the case when the total state is of the form $|000..0\rangle$ or $|111..1\rangle$, there is always two-qubit entanglement present in symmetric states. Note, however, that in the limit of n and k becoming large — no matter what their ratio may be — the value of the left-hand side approaches the value of the right-hand side and entanglement thus disappears. This is a very interesting property of symmetric states and we will be able to quantify it exactly in the next section.

An important point to make is that the two-point correlation function used in the calculation of the ODLRO in Eq. (4.7) is, in fact, just one of the 16 numbers we need for the full two-site density matrix (the independent number of real parameters is actually 15, because of normalization). In our simplified case of symmetric states in the η-pairing model, this off-diagonal element is equal to c. However, for the density matrix we still need to know a and b, and these numbers clearly affect the amount of entanglement. Imagine, for example, the situation where $a = b$. Then the condition for entanglement is that $a - c < 0$, which does not hold if $a \geq c$

and such a density matrix is certainly possible. So, the first lesson is that two-site entanglement is not the same as the existence of ODLRO, and therefore two-site entanglement is not relevant for superconductivity. This does not mean, of course, that there is no entanglement in the whole of the lattice. In the next section, it will be calculated exactly. We will determine the relative entropy of entanglement for all symmetric states and all their substrates. Through the approach, it will be able to extend the method of Wei *et al.* [58] and analyze many relationships between various subsets of symmetric states, including the amount of entanglement in any subset of qubits (or sites).

4.2.3. *Relative entropy of entanglement for symmetric states*

The symmetric states are very convenient for studying various features of multipartite entanglement simply because the relative entropy of entanglement can be computed for any reduced state including the total symmetric state for any n and k. It is expected that, because they possess a high degree of symmetry, they will also display a high degree of entanglement. It is precisely for this reason that they are suitable to allow the existence of entanglement at high temperatures. This will now be analyzed in detail.

We first introduce the relative entropy of entanglement. The relative entropy of entanglement measures the distance between a state and the nearest disentangled (separable) state. If $\mathcal{D}$ is the set of all disentangled states (i.e. states of the form $\sum_i p_i \rho_1^i \otimes \rho_2^i \cdots \otimes \rho_n^i$, where p_i is any probability distribution), the measure of entanglement for a state σ is then defined as

$$E(\sigma) := \min_{\rho \in \mathcal{D}} S(\sigma || \rho), \tag{4.13}$$

where $S(\sigma||\rho) = tr(\sigma \log \sigma - \sigma \log \rho)$ is the relative entropy between the two density matrices ρ and σ. In order to compute this measure for any state σ, we need to be able to find its closest disentangled state ρ. Finding this closest state is, in general, still an open problem, however, it has recently been solved for pure symmetric states by Wei *et al.* [58].

They showed that a convenient and intuitive way of writing the closest disentangled state to the symmetric state $|k, n-k\rangle$ is [58]:

$$\rho = \frac{1}{2\pi} \int_0^{2\pi} d\phi |\phi^{\otimes n}\rangle\langle\phi^{\otimes n}|, \tag{4.14}$$

where

$$|\phi^{\otimes n}\rangle = (\sqrt{k/n}|0\rangle + \sqrt{(n-k)/n}e^{i\phi}|1\rangle)^{\otimes n} \tag{4.15}$$

is the tensor product of n states each of which is a superposition of the states $|0\rangle$ and $|1\rangle$ with probabilities k/n and $1 - k/n$, respectively. This ρ was proved to achieve the minimum of the relative entropy by showing that it saturates an independently obtained lower bound. The relative entropy of entanglement of the total state is now easily computed. Since $\sigma = |k, n-k\rangle\langle k, n-k|$ is a pure state, $tr\sigma \log \sigma = 0$ and we only need to compute $-\langle k, n-k| \log \rho |k, n-k\rangle$, which is equal to

$$E(|k, n-k\rangle) = -\log \binom{n}{k} + k \log \frac{n}{k} + (n-k) \log \frac{n}{n-k}. \tag{4.16}$$

Note that entanglement is largest when $n = 2k$ as is intuitively expected (i.e. the largest number of terms is then present in the expansion of the state in terms of the computational basis states). Then, for large n, it can be seen that the amount of entanglement grows as

$$E(|n/2, n/2\rangle) \approx \frac{1}{2}(\log n + 2) \tag{4.17}$$

and so (in the leading order) entanglement grows logarithmically with the number of qubits in the state. To obtain this formula, we have used Sterling's approximation for the factorial

$$n! \approx 2.507 n^{n+1/2} e^{-n}. \tag{4.18}$$

Most results in this paper will asymptotically have the form $\alpha \log n + \beta$ where $\alpha > 0$ and β are constants that will usually be omitted as we only care about the general form of the behavior.

We now return to the question of different phases introduced between different elements of the superposition in the symmetric states. Let us consider states of the form

$$|1, n-1, \theta\rangle = |00..1\rangle + e^{i\theta}|00..10\rangle + e^{(n-1)i\theta}|10..0\rangle, \tag{4.19}$$

where we have $k = 1$ ones and $n - 1$ zeroes and θ is any phase. The simplest way of seeing that entanglement does not depend on the phase θ is to define a new basis at the mth site as $|\tilde{0}\rangle = |0\rangle, |\tilde{1}\rangle = \exp\{(m-1)\theta\}|1\rangle$. This way the phases have been absorbed by into the basis states and the resulting state is, in the tilde basis,

$$|1, n-1, \theta\rangle = |\tilde{0}\tilde{0}..\tilde{1}\rangle + |\tilde{0}\tilde{0}..\tilde{1}\tilde{0}\rangle + |\tilde{1}\tilde{0}..\tilde{0}\rangle. \tag{4.20}$$

The amount of entanglement must therefore be independent of any phase difference of the above type and this is, of course, true for symmetric states with any number of zeroes and ones. All considerations from this point onwards will therefore immediately apply to all these states with different phases.

We can also compute the two-site relative entropy of entanglement exactly. The closest disentangled state is in this case the same as in Eq. (4.14) with $n = 2$. In the computational basis, we have

$$\rho = \left(\frac{k}{n}\right)^2 |00\rangle\langle 00| + \left(\frac{n-k}{n}\right)^2 |11\rangle\langle 11| + \left(\frac{2k(n-k)}{n^2}\right) |\psi^+\rangle\langle\psi^+|. \tag{4.21}$$

That this is a minimum can be seen from the fact that the relative entropy of the state of two qubits is

$$\begin{aligned} S(\sigma||\rho) &= -S(\sigma) - \langle\psi^+|\log\rho|\psi^+\rangle - \langle 00|\log\rho|00\rangle - \langle 11|\log\rho|11\rangle \\ &\geq -S(\sigma) - \log\langle\psi^+|\rho|\psi^+\rangle - \log\langle 00|\rho|00\rangle - \log\langle 11|\rho|11\rangle, \end{aligned} \tag{4.22}$$

the inequality following from concavity of the log function. Suppose now that ρ's only non-zero elements are $\rho_{00} = \langle 00|\rho|00\rangle$, $\rho_{11} = \langle 11|\rho|11\rangle$ and $\rho_{++} = \langle\psi^+|\rho|\psi^+\rangle$. Given that it has to be separable, meaning that $2\sqrt{\rho_{00}\rho_{11}} \geq \rho_{++}$ (which follows from the Peres–Horodecki criterion), and that, at the same time, it has to be closest to σ, we can conclude that $\rho_{00} = k/n$. The other entries of ρ then follow.

To prove that ρ is the minimum in a rigorous fashion, we need to show that any variation of the type $(1-x)\rho + x\omega$ where ω is any separable state leads to a higher relative entropy (a method similar to [57]). Since relative entropy is a convex function, this means that

$$\frac{d}{dx} S(\sigma||(1-x)\rho + x\omega) \geq 0. \tag{4.23}$$

In fact, since relative entropy is convex in the second argument it is enough to assume that ω is just a product state.

For $a > 0$, $\log a = \int_0^\infty \frac{at-1}{a+t}\frac{dt}{1+t^2}$, and thus, for any positive operator A, $\log A = \int_0^\infty \frac{At-1}{A+t}\frac{dt}{1+t^2}$. Let $f(x,\omega) = S(\sigma||(1-x)\rho + x\omega)$. Then

$$\begin{aligned}\frac{\partial f}{\partial x}(0,\omega) &= -\lim_{x\to 0}\mathrm{Tr}\left\{\frac{\sigma(\log((1-x)\rho + x\omega) - \log\rho)}{x}\right\}\\ &= \mathrm{Tr}\left\{\left(\sigma\int_0^\infty (\rho+t)^{-1}(\rho-\omega)(\rho+t)^{-1}dt\right)\right\}\\ &= 1 - \int_0^\infty \mathrm{Tr}(\sigma(\rho+t)^{-1}\omega(\rho+t)^{-1})dt\\ &= 1 - \int_0^\infty \mathrm{Tr}((\rho+t)^{-1}\sigma(\rho+t)^{-1}\omega)dt. \end{aligned}\tag{4.24}$$

For our minimal guess ρ in Eq. (4.21) we can then write

$$\begin{aligned}\frac{\partial f}{\partial x}(0,\omega) - 1 &= -\mathrm{Tr}\left\{\omega\int_0^\infty (\rho+t)^{-1}\sigma(\rho+t)^{-1}dt\right\}\\ &= \frac{n}{n-1}\frac{k-1}{k}\langle 00|\omega|00\rangle + \frac{n}{n-1}\frac{n-k-1}{n-k}\langle 11|\omega|11\rangle\\ &\quad + \frac{n}{n-1}\langle\psi^+|\omega|\psi^+\rangle, \end{aligned}\tag{4.25}$$

where we have used the fact that $\int_0^\infty (p+t)^{-2}dt = p^{-1}$. Since the expression in the previous equation is always less than or equal to a unity if $\omega = |\alpha\beta\rangle\langle\alpha\beta|$ (i.e. a product state), it follows that

$$\left|\frac{\partial f}{\partial x}(0,\omega) - 1\right| \leq 1. \tag{4.26}$$

Thus it also follows that $\frac{\partial f}{\partial x}(0, |\alpha\beta\rangle\langle\alpha\beta|) \geq 0$. But any separable state can be written in the form $\rho = \sum_i r_i|\alpha^i\beta^i\rangle\langle\alpha^i\beta^i|$ and so

$$\frac{\partial f}{\partial x}(0,\rho) = \sum_i r_i\frac{\partial f}{\partial x}(0, |\alpha^i\beta^i\rangle\langle\alpha^i\beta^i|) \geq 0. \tag{4.27}$$

And this confirms that ρ is the minimum since the gradient is positive.

Therefore, the relative entropy of entanglement between any two sites is

$$\begin{aligned} E_{12} &= a\log a - b\log b - 2c\log 2c \\ &\quad - a\log\left(\frac{k}{n}\right)^2 - b\log\left(\frac{n-k}{n}\right)^2 - 2c\log\left(\frac{2k(n-k)}{n^2}\right) \\ &= \log\left(\frac{n}{n-1}\right) + \frac{k(k-1)}{n(n-1)}\log\left(\frac{k-1}{k}\right) \\ &\quad + \frac{(n-k)(n-k-1)}{n(n-1)}\log\left(\frac{n-k-1}{n-k}\right). \end{aligned} \tag{4.28}$$

We see that when $n, k, n-k \to \infty$, then $E_{12} \to 0$ as it should be from our discussion of the separability criterion. This can be thought of as one way of recovering the "quantum to classical" correspondence in the limit of large number of systems present in the state: locally, between any two sites, entanglement does vanish, although globally, and as will be seen in more detail, entanglement still persists.

Entanglement of any number of qubits, $l \leq k$, can also be calculated using the same method. The state after we trace out all but l qubits is given by

$$\sigma_l = \sum_{i=0}^{l} \binom{l}{l-i} \frac{\binom{n-l}{k-i}}{\binom{n}{k}} |i, l-i\rangle\langle i, l-i|. \tag{4.29}$$

The closest disentangled state is given by

$$\rho_l = \sum_{i=0}^{l} \binom{l}{i} \left(\frac{k}{n}\right)^{l-i} \left(\frac{n-k}{n}\right)^{i} |i, l-i\rangle\langle i, l-i|, \tag{4.30}$$

as can be shown by the above method. The relative entropy of entanglement is now given by

$$E_l = \sum_{i=0}^{l} \binom{l}{l-i} \frac{\binom{n-l}{k-i}}{\binom{n}{k}} \log\left\{\binom{l}{l-i} \frac{\binom{n-l}{k-i}}{\binom{n}{k}} \left(\frac{n}{k}\right)^{l-i} \left(\frac{n}{n-k}\right)^{i} \binom{l}{i}^{-1}\right\}. \tag{4.31}$$

This is a very interesting quantity as it allows us to speak about entanglement involving any number of qubits. What do we expect from it? We expect that entanglement grows exponentially with l, for a fixed total number of qubits, n. This can be confirmed using the Sterling formula. Note that entanglement grows at this rate even though the states we are talking about are mixed, since $n-l$ qubits have been traced out. Another way of seeing why entanglement grows exponentially with the number of qubits included for a total fixed number of qubits, is to look at the opposite regime. For any finite fixed l, we should have that in the large $n, k, n-k$ limit the amount of entanglement between l tends to zero. This decrease with larger and larger n happens at an exponential rate.

4.2.4. *Classical versus quantum correlations*

In this section, we try to investigate the relationship between classical and quantum correlations for symmetric states, and both in relation to the already introduced concept of ODLRO. First of all, it is clear that in the limit of $n \to \infty$ all bipartite (or two-site) entanglement disappears (this was seen both from the Peres–Horodecki criterion and from the direct computation of the relative entropy). In spite of this, the ODLRO still exists and the two quantities are therefore not related. In other words, two-site entanglement is not relevant for superconductivity. However the main point of this section is that the two-site classical correlations still survive in the limit of $n \to \infty$. In order to show this, let us, first of all, define bipartite classical correlations.

A quantum state can have zero amount of entanglement, but still have non-zero classical correlations. An example is the state $|00\rangle\langle 00| + |11\rangle\langle 11|$. Classical correlations between systems A and B in the state σ_{AB} can be defined as [70]

$$C_A(\sigma_{AB}) := \max_{A_i^\dagger A_i} S(\sigma_B) - \sum_i p_i S(\sigma_B^i) = \max_{A_i^\dagger A_i} \sum_i p_i S(\sigma_B^i \| \sigma_B), \quad (4.32)$$

where $\sigma_B^i = tr_A \sigma_{AB}^i$, $\sigma_{AB}^i = A_i \sigma_{AB} A_i^\dagger$, and $\sum_i A_i^\dagger A_i = 1$ is the most general measurement on system A. The same can be defined with the most general measurement performed on B, so that we obtain

$$C_B(\sigma_{AB}) := \max_{B_i^\dagger B_i} S(\sigma_A) - \sum_i p_i S(\sigma_A^i) = \max_{B_i^\dagger B_i} \sum_i p_i S(\sigma_A^i \| \sigma_A). \quad (4.33)$$

The physical motivation behind the above definition is the following: classical correlations between A and B tell us how much information we can obtain about A (B) by performing measurements in B (A). It is the (maximum) difference between the entropy of A (B) before and after the measurement on B (A) is performed. There is some evidence that $C_A = C_B$ [70], but this equality will not be relevant here.

Now, applying this measure of classical correlations to the two-site reduced density matrix from the overall symmetric state, ρ_{12}, we obtain

$$\begin{aligned} C &= -a\log a - b\log b - c\log c + \frac{1}{2}((a+c/2)\log(a+c/2) \\ &\quad + (b+c/2)\log(b+c/2)) \\ &= (r-2r^2)\log r + ((1-r)-2(1-r)^2)\log(1-r) \\ &\quad - 2r(1-r)\log 2r(1-r), \end{aligned} \tag{4.34}$$

where $r = k/n$ is the fraction of ones in the state (the so-called filling factor in any "Cooper pair" lattice model, including the η model). We now see that at half filling — when ODLRO is maximal — the classical two-site correlations also survive asymptotically since $C_A = C_B = 0.5$. Therefore, all the correlations between any two sites are here due to classical correlations.

Note, incidentally, that we cannot have the situation in which entanglement exists between two parties, while at the same time classical correlations vanish. Quantum correlations presuppose the existence of classical correlations. This, of course, relies on the fact that entanglement is defined in a reasonable way, namely that when we talk about two-site entanglement we must trace the other sites out. We are not allowed to perform measurements on other sites and condition the remaining entanglement on them.

Measurements that generate entanglement are, first of all, unrealistic for a macroscopic object which thermalizes very quickly. Even if we were to allow such measurements, then the state after them will still have classical correlations of at least the same magnitude as entanglement. So, it cannot be that entanglement is important for the issues of superconductivity, phase transitions, condensation, etc., and that classical correlations are not.

As an example, let us take the "maximum singlet fraction" in the two-site density matrix σ_{12} as our definition of entanglement. This is the maximum fraction of a maximally entangled state in the state σ_{12}, which is in this case equal to c, and this is the same as ODLRO. So, if the maximum singlet fraction is used to measure entanglement, then entanglement also persists in the thermodynamical limit. In fact, as will be shown later, this measure also survives when we mix symmetric states, because it is a linear measure. The maximum singlet fraction, however, is not a realistic measure of entanglement as it is not easily accessible experimentally, which is why we do not use it here.

In order to make our analysis more complete we also show how to calculate mutual information [55] for symmetric states. This quantity tells us about the total (quantum plus classical) correlations in a given state. Mutual information is equal to the relative entropy between the state itself and the product of individual qubit density matrices, obtained by tracing out all the other qubits. This product state is easily written down to be:

$$\rho_{prod} = \left(\frac{k}{n}|0\rangle\langle 0| + \frac{n-k}{n}|1\rangle\langle 1|\right)^{\otimes n}. \tag{4.35}$$

The mutual information is now given by

$$I(|k, n-k\rangle) = n\left(-\frac{k}{n}\log\frac{k}{n} - \frac{n-k}{n}\log\frac{n-k}{n}\right), \tag{4.36}$$

and this is basically just the sum of individual qubit entropies. Since the qubit entropy (the quantity in brackets in the above equation) is a finite quantity for a given ratio $r = k/n$, the total mutual information grows linearly with the number of qubits n. Furthermore, since entanglement grows as $\log n$, we conclude that classical correlations grow roughly as $n - \log n$ (for this conclusion to be exact, classical and quantum correlations as defined here would have to add up to mutual information; while this is true for some states [70], it is certainly not true in general).

The fact that classical correlations and mutual information survive the thermodynamical limit does not imply that there is no meaning left for entanglement when it comes to superconductivity and ODLRO. Only now, we must talk either about the bipartite entanglement between two clusters of sites (to be computed in the next section) or the multipartite entanglement between all sites. Since the overall state across all sites is pure in our considerations so far, this means that two-site non-vanishing classical correlations (or equivalently ODLRO) must imply entanglement between two clusters, each of which contains one of the sites and such that the union of the two clusters is the whole lattice. This simply must be the case, since, otherwise, if the clusters were not entangled, the total state would be a product of the states of individual clusters, and this means that even classical correlations would be zero, which is a contradiction. Furthermore, the fact that any two such clusters are entangled, must mean that the multipartite entanglement also exists, since this entanglement is by definition larger than any bipartite entanglement (as, for multipartite entanglement, we are looking for the closest separable state over all sites, rather than just over the two clusters). We now quantify these various relations a bit more precisely.

4.2.5. *Various other relations between entanglements*

In this section, we discuss some other results that can be derived from our knowledge of symmetric states so far. Some of the results will not necessarily be relevant directly for the main theme — high temperature entanglement — but that is a natural place to present them. The first important question to be addressed here is the following. Suppose we look at the entanglement between one qubits and the rest $n-1$ qubits in total and individually. We would expect that the total one-versus-rest entanglement is larger than the individual sum of the two-qubit entanglements. The logic behind this conclusion is that by looking at entanglements individually we always lose something from the total entanglement, i.e. the operation of tracing reduces entanglement. This translates into the following inequality:

$$(n-1)E_{12} \leq E_{1:(2,3..n)}, \tag{4.37}$$

where

$$E_{1:(2,3..n)} = -\frac{k}{n}\log\left(\frac{k}{n}\right) - \frac{n-k}{n}\log\left(\frac{n-k}{n}\right) \tag{4.38}$$

is basically the same as the entropy of every qubit in the symmetric state. We can prove this inequality by noting that it holds for $k = 1$ and $2k = n$ (the extreme points), and because of the monotonicity and continuity of both sides it has to hold in general.

The aforementioned inequality has a very important implication which shows that the bipartite entanglement in the symmetric state is always bounded from above by

$$E_{12} \leq \frac{E_{1:(2,3..n)}}{(n-1)} \leq \frac{1}{n-1} \approx \frac{1}{n} \tag{4.39}$$

the second inequality following from the fact that the entanglement between one qubit and the rest is equal to the entropy of that qubit and that can at most be $\log 2 = 1$. Therefore, while the total entanglement of the symmetric state increases with $\log n$, the two-qubit entanglement decreases as $1/n$. Here we see most directly how it is possible to have the emergence of (only) classical correlations between constituents even though globally entanglement increases. There are many other open questions related to this one. We can repeat the same calculation for any fixed number of qubits.

We can check if a cluster of qubits is, for example, more entangled to another cluster of qubits in total or if we add all the entanglements between their individual elements. Some of these may not be easy questions to answer in general.

We would now like to calculate the entanglement between l qubits and the remaining $n-l$ qubits. Since the whole state that we are now examining is pure, the relative entropy of entanglement is given by the entropy of the l qubits:

$$S_{12\ldots l} = -\sum_{i=0}^{l} \binom{l}{l-i} \frac{\binom{n-l}{k-i}}{\binom{n}{k}} \log \left\{ \binom{l}{l-i} \frac{\binom{n-l}{k-i}}{\binom{n}{k}} \right\}. \tag{4.40}$$

What are the properties of this expression when we take the various asymptotic limits? How is this quantity related to other entanglements calculated here? We expect that for the half filling, $n/k = 2$, and $n, l \to \infty$, the entropy becomes $\log l$, since we basically have a maximal mixture in the symmetric subspace of l qubits. This can be confirmed by a simple application of the Sterling approximation formula used before. The result is in agreement with the fact that total entanglement grows at the rate of the log of the number of qubits, since two cluster entanglement is a lower bound for the total entanglement in the state between all the qubits.

The last question we address is the relationship between the lower and higher order entanglement in the symmetric states. More precisely, the question is: if we add all the entanglements up to and including m qubits, is this quantity larger or smaller than the amount of entanglement of $m+1$ qubits? Mathematically, this translates into the following two possible inequalities:

$$\sum_{i=1}^{m} E_i \leq E_{m+1} \quad \text{or} \quad \sum_{i=1}^{m} E_i \geq E_{m+1} \tag{4.41}$$

where E_m is given in Eq. (4.31). We already know that for $n = 3$ and $k = 1, 2$, and $l = 3$ we have the equality in the above, namely $E_3 = E_1 + E_2$ [59]. From this result alone it is not clear which way to expect the inequality to be. Numerical examples show us that, in fact, both results are possible. If we check the inequality for $n = 100, k = 50$ and $l = 4$ for example, then the left-hand side is smaller than the right-hand side and the first inequality holds. For $n = 100, k = 50$ and $l = 30$, on the other hand, the left-hand side is larger than the right-hand side and the second inequality is satisfied. It is an interesting and open question to investigate the point of the cross-over when the two sides become equal to each other.

4.2.6. *Thermal entanglement and superconductivity*

There is a critical temperature beyond which any superconductor becomes a normal conductor. The basic idea behind computing this temperature according to BCS is the following. At a very low temperature, only the ground state of the system is populated and for a superconductor this state involves a collection of Cooper pairs with different momenta values around the Fermi surface. This state can be, somewhat loosely, thought of as a Cooper pair condensate, and it is this condensation that is the key to superconductivity.

It took initially a long time to understand how the pairs are formed, since electrons repel each other and therefore should not be bound together. The attraction is provided by electrons interacting with the positive ions left in the lattice. We can think of one electron moving and dragging along the lattice, which then pulls other electrons thereby providing the necessary attraction [22]. When the temperature starts to increase, the Cooper pairs start to break up, leading to the transition to the normal conductor. What this "breaking up" means is that higher than ground states start to get populated by electrons, and these are states where an electron is created with say momentum k and spin up, but no electron is created in the $-k$ momentum state.

From the BCS analysis this critical temperature can be calculated to be of the form [22]

$$T_c \approx \frac{\hbar\omega}{k} e^{-1/\lambda}, \tag{4.42}$$

where $\hbar\omega$ is the energy shell around the Fermi surface which is engaged in formation of Cooper pairs, k is the Boltzmann constant and λ is a parameter equal to the product of the electron density at the Fermi surface $N(0)$ and the effective electronic attractive coupling, V. The critical temperature formula is valid in the weak coupling regime where $\lambda = N(0)V \ll 1$.

The formula for the critical temperature is usually used for other mechanisms of electron pairing, and not just coupling via the phonon lattice modes as in the BCS model [22]. Importantly for us, the formula also features in models for explaining and designing high T_c superconducting materials. If the attraction, say between an electron and a hole, is of the order of Coulomb forces, $\hbar\omega \approx 1\,\mathrm{eV\,m}$, and for the weak coupling of, say, $\lambda = 0.2$, the critical temperature we obtain is 100 K. So if the material is below this temperature, it is then superconducting. Anything above 70–90 K is considered to be high temperature superconductivity, since it can be achieved by cooling with liquid nitrogen (which is a standard and easy method of cooling). What seems to be the mechanism behind high temperature superconductivity, is the fact that the energy gap between the ground superconducting, electron-pair state, and the excited states is large enough not to be easily excited as the temperature increases well beyond zero temperature. The exact way in which this is achieved is still an open question. In the models mentioned here the ground state is one of the symmetric states from the previous sections. Therefore, we can conclude that as long as we have high temperature superconductivity, the total state should also be macroscopically entangled. Superconduction and hence entanglement can currently exist at temperatures of about 160 K.

We would now like to explicitly calculate and show how entanglement disappears as the temperature increases for any model having the η pairing state as the ground state. For this, we need to be able to describe other states that would be mixed in with the symmetric η states as the temperature increases. They, of course, depend on the actual Hamiltonian. For instance, in the Hubbard model in [60], states of the type

$$\xi_a^\dagger|0\rangle = \sum_i c_{i,\downarrow}^\dagger c_{i+a,\uparrow}^\dagger|0\rangle \tag{4.43}$$

are important; here we create a spin singlet state but at sites separated by the distance $a \neq 0$. If we have $2k$ electrons in total, then $2k-2$ would be paired in the lowest energy state, and the remaining two electrons would not be. This would give us the state of the form

$$|\xi\rangle := \eta^{k-1}\xi_a^\dagger|0\rangle. \tag{4.44}$$

Note that this state is a symmetric combination of states which have $k-1$ electron pairs distributed among n sites and the last electron pair is in two different sites separated by the distance a. These two sites are different from the other $k-2$ sites due to Pauli's exclusion principle. Even higher states are obtained by having two electron pairs existing outside of the symmetric state and so on. The exact form of these, as noted before, depends on the exact form of the Hamiltonian. Even simple Hamiltonians are frequently very difficult to diagonalize and their eigenstates are still by and large unknown. Given this, it may be difficult to calculate the exact amount of entanglement when, at finite temperature, the ground state is mixed with higher energy states. Therefore, we make a simplifying assumption that, if the ground state is $|k, n-k\rangle$, the higher energy states can be written as $|k-1, n-k+1\rangle$, $|k-2, n-k+2\rangle$ and so on. All these in fact is assumed to be symmetric and we ignore the extra unpaired electrons as far as entanglement is concerned (they only contribute to the eigenvalue of energy as it were).

This assumption leads us to consider mixtures of symmetric states. The symmetric states will be mixed with probabilities in accordance to Boltzmann's exponential law, or the Fermi–Dirac law if we talk about η pairs. The distribution we use is immaterial for our argument. The total state, σ_T, is

$$\sigma_T = \sum_{k=0}^{n} p_k |\Psi(k,n)\rangle\langle\Psi(k,n)|, \tag{4.45}$$

where, in the case of η pairs, the probabilities are

$$p_i = \frac{1}{e^{E_i/kT}+1} \tag{4.46}$$

where p_i is the probability of occupying the ith energy level. The reduced two-site state can be calculated to be

$$\sigma_{12} = \sum_{k=0}^{n} p_k \sigma_{12}(k). \tag{4.47}$$

The condition for inseparability now becomes

$$\sum_{k,l} p_k p_l k(n-l)\{(n-k)l - (k-1)(n-l-1)\} > 0. \tag{4.48}$$

We see that the thermal averaging is in a sense inconsequential for the existence of entanglement as the factors $p_k p_l$ are probabilities and are always non-negative. For inequality to hold (i.e. to have non-zero bipartite entanglement present) we need that $1 \leq k, l \leq n-1$. This is the same condition as before when the total state was pure. Thus, surprisingly, the condition for inseparability is completely independent of temperature (although, two-site states do become separable in the macroscopic limit even at zero temperature, as noted before).

We now look at the entanglement of the symmetric mixed state as a whole. Can we still calculate the relative entropy of entanglement in the finite temperature limit? This is in general very difficult to do for multiparty mixed states, and some partial methods for upper bounds have only been presented recently [34]. We conjecture that the closest disentangled state is now presumably the thermal average of the closest disentangled states for individual k's (this, I believe, is the same as the conjecture in [58], for which Wei *et al.* have offered a great deal of "circumstantial evidence"; for example, closest separable states have to possess the same symmetry as the entangled states for which they minimize the relative entropy [71]). We conjecture that this bound is exact and that this can be proven using methods for calculating two-site entanglement, although it cannot be proven here. Even if the conjecture is not true, our approach at least gives us a very good upper bound which is sufficient to show how total entanglement vanishes as T becomes high.

The relative entropy of entanglement between these two states is given by (the right-hand side of the inequality)

$$E(\sigma_T) \leq \sum_k p_k \log p_k - \sum_k p_k \langle \Psi(k,n)| \log \left(\sum_l p_l \rho_l \right) |\Psi(k,n)\rangle, \quad (4.49)$$

where ρ_l is the closest disentangled state to the pure symmetric state containing l ones and $n-l$ zeroes. We have already seen that

$$\rho_l = \sum_{i=1}^{l} \binom{l}{i} \left(\frac{k}{n}\right)^{l-i} \left(\frac{n-k}{n}\right)^{i} |\Psi(l,n)\rangle\langle\Psi(l,n)|, \quad (4.50)$$

so that

$$\begin{aligned} E(\sigma_T) &\leq \sum_k p_k \log p_k \\ &\quad - \sum_k p_k \langle \Psi(k,n)| \log \left\{ \sum_l p_l \sum_{i=1}^{l} \binom{l}{i} \left(\frac{k}{n}\right)^{l-i} \left(\frac{n-k}{n}\right)^{i} \right\} \\ &\quad \times |\Psi(l,n)\rangle\langle\Psi(l,n)|\Psi(k,n)\rangle \\ &= -\sum_k p_k \log \sum_{i=1}^{k} \binom{k}{i} \left(\frac{k}{n}\right)^{k-i} \left(\frac{n-k}{n}\right)^{i}. \end{aligned} \quad (4.51)$$

The interesting conclusion here is the following. Suppose that we are at a high temperature and that all symmetric states are equally likely, meaning that $p_k = 1/(n+1)$ for all values k (basically, our state is an equal mixture of all symmetric states). This is, of course, in an approximation to the true density matrix, but it becomes more and more accurate with the increase of temperature and it ceases to be so when states other than symmetric become mixed in. The (upper bound to) entanglement is then given by

$$E(\sigma_{T\to\infty}) \leq \frac{1}{n+1} \log \sum_{i=1}^{k} \binom{k}{i} \left(\frac{k}{n}\right)^{k-i} \left(\frac{n-k}{n}\right)^{i}. \tag{4.52}$$

The fraction inside the log tends to n^2 as n becomes large, so that entanglement scales as $\log n/n$. This is to be expected as entanglement grows as $\log n$ with n, but the mixedness grows linearly with the number of state involved, $n+1$. Therefore, in the thermodynamical limit, the overall mixed state entanglement also disappears. This has to eventually happen, of course, if we believe that entanglement is intimately linked with superconductivity and superconductivity also vanishes at sufficiently high temperatures.

One kind of entanglement that we can say survives the thermodynamical high temperature limit is the average of entanglements of individual symmetric states. This average entanglement is given by

$$E_{\text{avr}} = \sum_k p_k E(|k, n-k\rangle) = \frac{1}{2} \sum_k p_k \log \frac{k(n-k)}{n}. \tag{4.53}$$

Note that if all probabilities go as $1/n$ — i.e. the symmetric state is maximally mixed, then the entanglement scales as $\log n$ (the same as pure state at half filling). This is expected, as there are $n+1$ states and each one has entanglement proportional to $\log n$ and, so, on average, entanglement also goes as $\log n$. However, this average entanglement, as we argued before is not a good measure as it requires us to be able to address the symmetric states individually and discriminate them from each other. This is not just difficult in practice, but is in fact frequently even impossible in principle.

It is interesting to note that the ODLRO does survive the mixing of symmetric states. Even when we have an equal mixture of all symmetric states, the average ODLRO is given by

$$\frac{1}{n+1} \sum_{k=0}^{n} \frac{k}{n} \frac{(n-k)}{n} = \frac{1}{2} - \frac{1}{6} \frac{2n+1}{n} \to \frac{1}{6}, \tag{4.54}$$

where the arrow indicates the convergence when n is large. Of course, at sufficiently high temperatures the system will leave the subspace of symmetric states and other states will also start to contribute.

This eventually does lead to vanishing of ODLRO, but the total entanglement and ODLRO may still disappear at different temperatures. To calculate this exactly, we would need a much more detailed model and a more extensive and careful calculation which lie outside of the scope of the current literature. (Note: the same conclusions hold for the maximum singlet fraction in the two-site density matrix which also survives the mixing in the thermodynamical limit; this is, unfortunately and as pointed out before, not a suitable measure of entanglement in our setting.)

Here, we conclude by showing that total correlations — quantum and classical — as quantified by the mutual information [55] can also easily be calculated for thermal mixtures of symmetric states. Let us assume again that the symmetric states are maximally mixed and each appears with the probability $1/(n+1)$. Then the mutual information is given by

$$I = \frac{n}{n+1}\sum_k \left(-\frac{k}{n}\log\frac{k}{n} - \frac{n-k}{n}\log\frac{n-k}{n}\right) - \log(n+1). \tag{4.55}$$

For large n this expression reduces to

$$I \to n - \log n. \tag{4.56}$$

Since we know that thermal entanglement disappears in this limit, it is natural that the mutual information is equal to the classical correlations and this then coincides with the conclusion following (4.36).

4.2.7. *D-dimensional symmetric states*

Extensions of all the considerations in this paper to D-dimensions are seen to be very straightforward (similar generalizations to higher dimension symmetric states were also considered in [58]). We should actually be able to reproduce all the above results in the generalized form, such that instead of qubits we have qutrits, and so on. The generic symmetric state would now be written as

$$|n_1, n_2, \ldots, n_d\rangle \tag{4.57}$$

and it would be a totally symmetrized state of n_1 states $|1\rangle$, n_2 states $|2\rangle$ and so on (it is also realistic to assume that the total number of particles is conserved). This could, for example, represent higher spin fermions which can occupy different lattice sites. The closest state to the one in Eq. (4.57) in terms of the relative entropy is a mixture of the states of the type

$$(\sqrt{n_1/N}|1\rangle + e^{i\phi}\sqrt{n_2/N}|2\rangle + \cdots + e^{i(d-1)\phi}\sqrt{n_d/N}|d\rangle)^{\otimes n}, \tag{4.58}$$

with the phase ϕ completely randomized as before. Knowledge of this closest state allows us to compute the relative entropy of entanglement of any number of subsystems of this system. All other results follow in exactly the same way. Nothing fundamental is changed in higher dimensions and we leave them to be done by readers more.

4.2.8. *Section remarks*

In this section, the η pairing mechanism is analyzed which leads to eigenstates of the Hubbard and similar models used in explaining high temperature superconductivity. It is shown that they correspond to multi-qubit symmetric states, where the qubit is made up of an empty and a full site (two-electron spin singlet state). It is also shown how to calculate entanglement and classical correlations for such states. For pure states, entanglement of the total state increases at the rate $\log n$ with the number of qubits n, while two-site entanglement vanishes at the rate $1/n$. The two-site classical correlations, on the other hand, persist in the thermodynamical limit. So, the ODLRO can be associated for pure states with total entanglement or two-site classical correlations, but not with the two-site entanglement. It is also demonstrated that the total entanglement for maximally mixed symmetric states disappears at the rate $(\log n)/n$. Various mutual information measures, which quantify the total amount of correlations in a given state, are also computed and shown to be consistent with the calculations of classical and quantum correlations.

There are many interesting issues raised through the discussion. Even if a consensus is reached on the correct model for high T_c superconductivity, and this is shown to contain multipartite electron entanglement — which we have argued already — we are still left with the question of being able to extract and use this entanglement. At present there are no methods of extraction. Perhaps we can somehow extract electrons from the superconductor and then use them for quantum teleportation or other forms of quantum information processing.

It is presently believed that in order to perform a reliable and scalable quantum computation we may need to be at very low temperatures, but the existence of high temperature macroscopic entanglement may just challenge this dogma. Be that as it may, the argument in favor of the existence of high temperature entanglement does show that entanglement may be much more ubiquitous than is presently thought. This may force us to push the boundary between the classical and the quantum world towards taking seriously the concept that quantum mechanics is indeed universal and should be applied at all levels of complexity, independently of the number or, indeed, nature of particles involved.

4.3. The Meissner Effect and Massive Particles as Witnesses of Macroscopic Entanglement

In the recent investigation, it is argued that high temperature macroscopic entanglement is possible and linked it to high T_c superconductivity [34]. The relationship between the notion of long-range off-diagonal order (ODLRO) [23] in a state and the existence of bipartite and multipartite entanglement in the same state has been discussed. Here, we intend to extend this line of thought and show that the said multipartite entanglement implies two typical superconducting effects: the exclusion of the magnetic field from a simply connected superconductor (the Meissner effect) and the quantization of flux in multiply connected regions. Although we will use the Hubbard related models to aid the discussion, our results hold for any model which displays ODLRO, i.e. normal superconductors and superfluids. They also conclude with a speculation, maintaining that if Higgs bosons are found to exist, and if they are responsible for mass generation through the symmetry breaking mechanism, then they must also be entangled. The reason is that the condensation of Higgs bosons is understood to be the most likely mechanism for mass generation in local gauge field theories (i.e. the Standard Model).

4.3.1. *Setting the scene of η-paring states*

Here, we start with recapitulation of the η-paring superconducting states which is the basic setting for our discussion. The model we analyze consists of a number of lattice sites, each of which can be occupied by fermions of spin up or spin down. Since fermions obey the Pauli exclusion principle, we can have at most two fermions attached to one and the same site. Let us introduce fermion creation and annihilation operators, $c_{i,s}^{\dagger}$ and $c_{i,s}$ respectively, where the subscript i refers to the ith lattice site and s refers for the value of the spin, $\uparrow$ or $\downarrow$. The c operators satisfy the anticommutation relations: $\{c_{i,s}, c^{\dagger}_{j,t}\} = \delta_{ij}\delta_{s,t}$, and c's and $c^{\dagger}$'s anticommute as usual. (Some general features of fermionic entanglement were analyzed in [64, 72, 73]).

We only need to assume that our model has the interaction which favors formation of Cooper pairs of fermions of opposite spin at each site — these states are known as η states [60] and will be discussed below. The actual Hamiltonian is not relevant for our present purposes. It suffices to say that the η states are eigenstates of the Hubbard and related models relevant for superconductivity [60, 67].

As we already seen, the η-paring superconducting model can be formulated as follows. Suppose that there are n sites and suppose, further, that we introduce an operator

$$\eta^{\dagger} = \sum_{i=1}^{n} c^{\dagger}_{i,\uparrow} c^{\dagger}_{i,\downarrow} \tag{4.59}$$

that creates a coherent superposition of a Cooper pair in each of the lattice sites. This $\eta^{\dagger}$ operator can be applied to the vacuum a number of times, each time creating a new coherent superposition. However, the number of applications, k, cannot exceed the number of sites, n, since we cannot have more than one pair per site due to the exclusion principle. We now introduce the following basis:

$$|k, n-k\rangle := \binom{n}{k}^{-1/2} (\eta^{\dagger})^{k} |0\rangle, \tag{4.60}$$

where the factor in front is just the necessary normalization. Here, the vacuum state $|0\rangle$ is annihilated by all c operators, $c_{i,s}|0\rangle = 0$. We note that the originally defined η operators can also have phase factors dependent on the location of the site on the lattice, like so $\eta_k = \sum_n e^{ikn} c^{\dagger}_{n,\uparrow} c^{\dagger}_{n,\downarrow}$. All the states generated with any η_k from the vacuum have the same amount of entanglement so that the extra phases will be ignored in the rest of the paper. However, the phase must be chosen, for otherwise, if we average over all possible phases, the resulting state is no longer entangled. Choosing a phase amounts to "symmetry breaking" and we will have more to say about it below.

We can think of the η states in the following way [34]. Suppose that $k = 2$. Then this means that we will be creating two η-pairs in total, but they cannot be created in the same lattice site. The state $|2, n-2\rangle$ is therefore a symmetric superposition of all combinations of creating two pairs at two different sites. Let us, for the moment, use the label 0 when the site is unoccupied and 1 when it is occupied. Then $|2, n-2\rangle = (|00...11\rangle + ...|11...000\rangle)/\sqrt{\binom{n}{2}}$, i.e. the state is an equal superposition of states containing 2 states $|1\rangle$ and $n-2$ states $|0\rangle$. These states, due to their high degree of symmetry, are much easier to handle than general arbitrary superpositions and we can compute entanglement for them between any number of sites [55]. In this description each site effectively holds one quantum bit, whose 0 signifies that the site is empty and 1 signifies that the site is full.

4.3.2. *Flux quantization*

Imagine superconducting material that has the region not simply connected but there is a hole in the middle pierced through by a magnetic field (there could be more than one hole and the same conclusion will hold for each of them). Then its flux must be quantized. This follows immediately from Eq. (3.122), since now not all paths are allowed. Namely, the field is now confined to the region where electrons cannot go, so that

$$\frac{2e}{\hbar c}\int\int BdS = \frac{2e}{\hbar c}\Phi_c \neq 0. \tag{4.61}$$

Therefore, we must have that $\frac{2e}{\hbar c}\Phi_c = 2n\pi$, and so the flux is quantized in units of $\hbar c/2e$:

$$\Phi_c = n\frac{\hbar c}{2e}. \tag{4.62}$$

This is the flux quantization effect. Note that the denominator contains twice the electron charge and this is a consequence of electrons forming Cooper pairs.

The flux quantization that we have just derived lies behind the persistent flow of electrical current in a superconductor. The flux is a consequence of the flowing current and any (continuous) dissipation cannot change the current continuously as the flux is discrete. Therefore, the current persists indefinitely.

Finally, if there is no ODLRO, meaning that as $n \to \infty$ we have that $c \to 0$ in Eq. (4.9), then neither of the above two effects follow. Any phase now gained upon exchange of electrons as described above will not be reflected in the two-site state and therefore we cannot argue that this phase has to have a special value. Therefore, bipartite entanglement is necessary for the Meissner effect and flux quantization. Note briefly that the converse is not true. Not all entanglement will lead to superconductivity. For example, look at the state of two sites of the form: $|00\rangle + |11\rangle$. When we exchange the pairs we get no extra phase in the second ket (because they have opposite signs and so their product equals identity), so that the state remains the same. Therefore, any magnetic field is allowed to permeate such a state. This is why the ODLRO concerns the coherences between states $|01\rangle$ and $|10\rangle$.

4.3.3. *Mass from entanglement between Higgs bosons*

Here, we would now like to discuss about the Higgs mechanism as the main explanation for the appearance of mass in local gauge field theories (see [75] for a comprehensive introduction). In the modern field theory, gauge invariance of the Hamiltonian (or Lagrangian, which is more typically used) is invoked to explain the appearance of fields and their bosonic mediators.

By "gauge transformation" we mean a transformation that acts on the wave function (or the field, more precisely) in the following manner $|\Psi(x,t)\rangle \rightarrow e^{\theta(x,t)}|\Psi(x,t)\rangle$, where $\theta(x,t)$ is just a phase that depends both on space and time (i.e. it is local). In order for the Hamiltonian to remain invariant under this local change, we need to introduce an extra (vector) field, whose features exactly cancel out the effects of the local phase change. The necessary field turns out to be the electromagnetic field and its bosons are, of course, photons. The important point is that if we are to derive other forces from local gauge invariance (this requires phases that are non-commuting — i.e. matrices, but the concept is the same), the resulting bosons will always be massless. This result is intuitively clear: the local phase change has to be matched between arbitrary points and times and this therefore requires an infinite range force. The mediators of infinite range forces have to be massless. So, it appears that local gauge invariance cannot explain forces whose mediators are massive.

A solution to this problem was found by Higgs [76] (and a number of other people, but Higgs was most prominent). The idea is that in addition to specifying the Hamiltonian of the fields, we also need to specify their actual physical state. This state need not possess the same symmetry properties of the Hamiltonian (hence the phrase "symmetry breaking" [75]). Suppose now that our local gauge invariance leads to several interacting massless fields. Suppose also, that one of the fields — known as the Higgs field — has condensed. In a mechanism that is completely analogous to the Meissner effect, the other fields will now be "expelled" from the Higgs field and will therefore become short-range. In other words, their mediators will become massive. The condensation of the Higgs field therefore provides a mechanism to maintain local gauge invariance and have massive gauge fields at the same time, thereby circumventing previously mentioned limitation (the Higgs boson also acquires a mass in this process). Whether this is the correct way of explaining the origin of mass in the universe is still unclear as the search for Higgs bosons has so far been fruitless. However, one conclusion we can draw with more confidence (following this paper) is that if the Higgs field exists, then its bosons must be entangled. The reason is that the ODLRO, necessary for condensation, also implies existence of entanglement, and this is also true for Higgs condensation. The (obvious) fact that there are massive objects in the universe would then be an entanglement witness of the purely quantum correlations in the underlying Higgs field.

Here we have to exercise some caution. The entanglement in the Higgs field is something that we refer to as "continuous variable entanglement", as opposed to the discrete degrees of freedom of the Hubbard model above, and this quantity can become infinite. However, this is not a serious problem because with enough care this infinity can always be controlled.

We now show in a very simple example the connection between mass and entanglement that is meant to substantiate the above discussion, but is by no means a proof of it. Suppose we have a massive bosonic free field, ϕ, with the usual Lagrangian density $1/2((\partial_\mu \phi)^2 + m^2\phi^2)$ (this is in $1+1$ dimensions), where m is the (fixed finite) mass. This is an infinite continuous system and we divide it into two halfs (arbitrarily). Tracing one part out and computing the von Neumann entropy of the remaining part results in the entropy of entanglement of $E \approx \ln 1/m^2a^2$ [77], where a is some cutoff used to avoid the ultraviolet infinity (this divergence may also be avoided by using the relative entropy with respect to some coarse graining [78], much in the same way as Gibbs did in classical statistical mechanics). The amount of entanglement clearly depends on the mass which could therefore be said to witness it.

4.3.4. *Spin entanglement*

We would like to point out an interesting curiosity that clarifies the notion of entanglement we have analyzed in the current discussion. Namely, as we mentioned before, the relevant entanglement for superconductivity (and Higgs bosons) is the spatial entanglement between numbers of electrons at different space points. What about the entanglement between the spin degrees of freedom? If the electrons occupy the same site, then they have to be anticorrelated (in the singlet state) because of Pauli's exclusion principle. If the electrons are on different sites, then they are not spin correlated in the η state (in the BCS model they would be, for a sufficiently small distance [79]). This is because if we measure an electron in one site and then in another all four possibilities for their internal states are equally likely and so there can be no spin correlation present.

There is a limit, however, in which the spin entanglement becomes relevant. This is when the interaction between sites dominates the hopping amplitude in the Hubbard model and in the state where we have one electron per site. Here there is no ODLRO and the state is not superconducting. However, the effective interaction between electrons at neighboring sites is now of the Heisenberg type, as electrons can still exchange their locations and could at the same time have opposite spins. Therefore, at low temperatures there would be some two-site spin entanglement present in the model, which is albeit not important for superconductivity. The way of witnessing entanglement of a spin under the influence of distributed external field will be further discussed in the next section.

4.3.5. *Superconducting qubit controlled by geometric phase*

The superconducting material contains large number of paired electrons, called cooper pair, and the pair behaves like bosonic particles with single excitation. The excitation can be treated as a logical quantum bit similarly to binary number for information coding. Using the basic unit, qubit, it has been proposed that quantum computation based upon non-abelian phases is possible [80, 81].

The quest for large-scale integrability has stimulated increasing interest in superconducting nanocircuits as possible candidates for the implementation of a quantum computer [82]. Mesoscopic Josephson junctions can be prepared in a controlled superposition of charge states and the coherent time evolution in a Josephson charge qubit has been observed. That are the first important experimental steps towards the implementation of a solid-state quantum computer. Later, it is further demonstrated that the controlled accumulation of a geometric phase, Berry phase, in a superconducting junctions is possible as it can be used for controlled qubit [84].

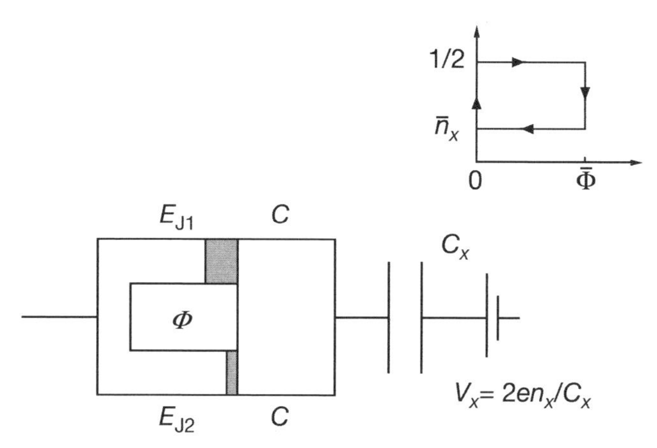

Fig. 4.1. The systems consist of a superconducting electron box formed by an asymmetric SQID, pierced by magnetic flux Φ, and with an applied gated voltage V_x. The offset charge n_x defined is proportional to V_x. The device operates in the charge regime, that is $E_{J1}, E_{J2} \ll E_{\text{ch}}$. The cyclic evolution of the Hamiltonian is obtained by changing the gate voltage and the magnetic flux along the path shown in the inset.

The simple description for the scheme at the level of principle has been proposed in [83]. The setup for the basic unit is considered as shown in Fig. 4.1. It consists of a superconducting electron box formed by asymmetric SQUID, pieced by a magnetic flux Φ and with an applied gate voltage V_x. The device operates in the charging regime, that is, the Josephson coupling $E_{\text{J1(2)}}$ of the junctions are much smaller than the charging energy E_{ch}. We further assume that the temperature is much lower than $E_{J1(2)}$. The Hamiltonian is [85]

$$H = E_{\text{ch}}(n - n_x)^2 - E_j(\Phi)\cos(\theta - \alpha), \tag{4.63}$$

$$E_j(\Phi) = \sqrt{(E_{J1} - E_{J2})^2 + 4E_{J1}E_{J2}\cos\left(\pi\frac{\Phi}{\Phi_0}\right)}, \tag{4.64}$$

where $\tan\alpha = \tan(\pi\frac{\Phi}{\Phi_0})(E_{J1} - E_{J2})/(E_{J1} + E_{J2})$ and $\Phi_0 = h/2e$ is the superconducting quantum of flux. The phase difference across the junction θ and the number of Cooper pairs n are canonically conjugate variables $[\theta, n] = i$. The parameters of the hamiltonian can be controlled. The offset charge $2en_x$ can be tuned by changing V_x in Fig. 4.1 and the coupling $E_J(\Phi)$ depends on Φ. The use of an asymmetric SQUID permits control of the phase shift $\alpha(\Phi)$ as well.

At temperatures much lower than E_{ch}, if n_x varies around the value $1/2$, only the two charge eigenstates $n = 0, 1$ are important. The constitute the basis $\{|0\rangle, |1\rangle\}$ of the computational Hilbert space of the qubit. The effective Hamiltonian is obtained by projecting Eq. (4.63) on the computational Hilbert space, and read $H_B = -(1/2)\mathbf{B} \cdot \sigma$, where we have defined the fictitious field $B = (E_J \cos\alpha, -E_J \sin\alpha,\ E_{\text{ch}}(1 - 2n_x))$ and σ are the Pauli matrices.

The system thus behaves like an artificial spin in a magnetic field. Charging couples the system to B_z whereas the Josephson term determines the projection in the xy plane. By changing V_x and Φ, the qubit Hamiltonian H_B describes a cylindroid in the parameter space $\mathbf{B}$. The presence of higher charge states leads to quantum leakage. This effect can be minimized by a fine-funding of the parameters of the devices [86].

The very recent research in superconducting quantum computing is conducted by glamorous commercial companies as like Google [87], IBM [88] and Intel [89]. As an reliable qubit array, IBM announced that up to nine fully controllable qubits are demonstrated in a 1D array, up to sixteen in a 2D architecture at this year, 2017.

4.3.6. *Section remarks*

In one of our previous discussion [34], we argued that macroscopic entanglement exists at high temperatures and is related to high temperature superconductivity. In this section, we showed that the consequences of that entanglement are the standard features of superconductors: the Meissner effect and flux quantization. Therefore any experiments confirming these two effects are also automatically offering evidence for macroscopic entanglement.

We have speculated that if the Higgs mechanism for mass generation is proven to be correct, then the resulting Higgs bosons will be found to be entangled. Be that as it may, one question remains open, both for superconductors, or for any other more general field. Can we extract this existing entanglement and use it for information processing? This would be very useful in practice, and it would seem that natural macroscopic entanglement could offer an infinite amount of quantum non-locality for genuine quantum information processing. This is the subject of an ongoing research.

4.4. Quantum Instability in a Quasi-Long-Range Ordered Phase

Strongly interacting many-body systems display a variety of quantum phase transitions (QPT) at absolute zero temperature [142]. Quantum criticality is a property of the ground state of the system, ultimately arising from a reshuffling of the system's energy spectrum when control parameters are varied. In many cases, this is accompanied by a symmetry breaking and by the appearance of an order parameter indicating that the macroscopic order is reached. Such kind of criticality is characterized by long-range correlations. There also exist phase transitions with vanishing local order parameter. A typical example is the Berezinskii–Kosterlitz–Thouless (BKT) transition, occurring, e.g. in a system of planar classical spins at a finite temperature [27]. The transition is characterized by a "distortion" of the spatial spin configuration of topological nature, giving rise to quasi-long-range order in the system, whose correlation functions display a power-law decay (see [143] for a review).

The prospect of practical applications, such as quantum information processing, has led to an intense activity aimed at a direct inspection of the ground state properties of quantum critical systems. The purpose of the present paper is to analyze the ground state of a system characterized by quasi-long-range order. Specifically, we focus on the spin $1/2$ XX model in external magnetic field in one dimension. According to the general equivalence between the classical-$d+1$ and quantum-d criticality [144], this model displays a QPT from a polarized to a BKT critical phase, with power-law-decaying correlation functions. We will investigate the fidelity and the entanglement encoded in the state as sensitive probes to the QPT [145, 146]. Moreover, the build-up of quasi-long-range order in the thermodynamic limit is traced back to the finite size behavior of two-point correlations and pairwise entanglement, and we show that, as the magnetic field is varied, the system undergoes a sequence of energy level crossings, leading to ground state instability. Indeed, the fundamental level discontinuously changes its quantum numbers by jumping from one spin sector of the Hilbert space to another. This is reflected in the correlation functions as well as in the ground state entanglement, displaying peculiar jumps. Similar kind of instability in a different model, the Dicke model, has been observed in [147]. The number of crossings grows (and their effect weakens) with the size of the chain. As $N \to \infty$, the crossing points become infinitesimally dense and continuous within a sharply defined critical region.

4.4.1. *The ground state of the XX model*

We consider N spin 1/2 particles on a line, coupled by nearest neighbor XX interaction, with Hamiltonian

$$H = -J\left[\sum_{i=1}^{N}\frac{1}{2}(\sigma_i^x\sigma_{i+1}^x + \sigma_i^y\sigma_{i+1}^y) + B\sigma_i^z\right] \tag{4.65}$$

where we have taken the exchange interaction J as the energy unit. In the thermodynamic limit, the system undergoes a first-order transition from a fully polarized to a critical phase with quasi-long-range order [143]. The model with periodic boundary condition has been firstly solved by Katsura [116] and recently discussed at finite size in [148]. To avoid unnecessary complications, we assume open boundaries, with $\sigma_{N+1} = 0$. Then, the Hamiltonian can be diagonalized via the Jordan–Wigner (JW) and Fourier transformations (FT). With the introduction of the following fermion operators

$$d_k = \sqrt{\frac{2}{N+1}}\sum_{l=1}^{N}\sin\left(\frac{\pi kl}{N+1}\right)\bigotimes_{m=1}^{l-1}\sigma_m^z\sigma_l^-, \tag{4.66}$$

the Hamiltonian takes the form, $H = \sum_{k=1}^{N}\Lambda_k d_k^\dagger d_k + NB\mathbb{1}$ where $\Lambda_k = -2B + 2\cos[(\pi k)/(N+1)]$. The 2^N eigenenergies are $\epsilon_i \equiv \sum_{k=1}^{N}\Lambda_k\alpha_k^{(i)} + NB$, where the corresponding eigenstates are $|\psi_i\rangle = \Pi_{k=1}^{N}(d_k^\dagger)^{\alpha_k^{(i)}}|\Omega\rangle$, where $\alpha_k^{(i)} = \langle\psi_i|d_k^\dagger d_k|\psi_i\rangle \in \{0,1\}$. The state $|\Omega\rangle$ is the fermion vacuum: $d_k|\Omega\rangle = 0\,\forall k$.

The ground state energy varies as a function of B. The different ground states can be classified in terms of the number of level crossings occurring in the system as B changes. Specifically, when $B > 1$, $\Lambda_k < 0$ for any k and, thus, the lowest eigenvalue is obtained by taking the state with $\alpha_k = 1\,\forall k$. The ground state energy remains $\epsilon_g^0 = -NB$ as long as $B > \cos[\pi/(N+1)]$. For these values of B no level crossing occurs; the ground state is $|\psi_g^0\rangle = \prod_{l=1}^{N} d_l^\dagger|\Omega\rangle$. The first crossing occurs at $B = \cos[\pi/(N+1)] \equiv B_1$. For $\cos[2\pi/(N+1)] < B < \cos[\pi/(N+1)]$ all of the Λ_k are negative except for Λ_1. Thus, the ground state energy is obtained by *subtracting* its positive contribution: $\epsilon_g^1 = \epsilon_g^0 - \Lambda_1$. The corresponding eigenstate is $|\psi_g^1\rangle = d_1|\psi_g^0\rangle$. Letting $B_k = \cos[k\pi/(N+1)]$ and defining the kth region, $B_{k+1} < B < B_k$, we can iterate the procedure above to find the ground state $|\psi_g^k\rangle = d_k d_{k-1} \cdots d_1 |\psi_g^0\rangle = \prod_{l=k+1}^N d_l^\dagger |\Omega\rangle$ and its energy as

$$\epsilon_g^k = -(N-2k)B - 2\sum_{l=1}^{k} \cos\left(\frac{\pi l}{N+1}\right), \tag{4.67}$$

where k represents the number of crossings, $1 \leq k \leq N$. For $B < B_N$, no other intersection occurs and the ground state is simply given by the fermion vacuum state. The energy crossings are plotted versus B in Fig. 4.3(a).

To investigate the structure of the ground state and its entanglement content, we rewrite the state in the spin basis, using the eigenstates of σ_i^z. The ground state is obtained by applying the inverse of JW and FT, in reverse order. For zero-level crossing the ground state is $|\varphi_g^0\rangle = |\uparrow\rangle^{\otimes N}$, which is separable. After the first crossing, it becomes $|\varphi_g^1\rangle = [\sum_{l=1}^N S_l^1 (\prod_{m=1}^{l-1} \sigma_m^z \sigma_l^-)] |\varphi_g^0\rangle$, where $S_l^k \equiv \sqrt{2/(N+1)} \sin[(\pi k l)/(N+1)]$. This is an entangled state, given by a symmetric superposition of all possible permutations of kets with just one spin flip. After k level crossings, the ground state is given by $|\varphi_g^k\rangle = \prod_{k'=1}^k [\sum_{l=1}^N S_l^{k'} (\prod_{m=1}^{l-1} \sigma_m^z \sigma_l^-)] |\uparrow\rangle^{\otimes N}$. Explicitly

$$|\varphi_g^k\rangle = \left(\frac{2}{N+1}\right)^{\frac{k}{2}} \sum_{l_1 < l_2 < \cdots < l_k} C_{l_1 l_2 \cdots l_k} |l_1, l_2, \ldots, l_k\rangle \tag{4.68}$$

where $|l_1, l_2, \ldots, l_k\rangle$ is the state with flipped spins at sites $l_1, l_2, \ldots, l_k$. The amplitudes in the state (4.68) are given by $C_{l_1 l_2 \ldots l_k} = \sum_P (-1)^P \sin(\frac{\pi P(1) l_1}{N+1}) \sin(\frac{\pi P(2) l_2}{N+1}) \cdots \sin(\frac{\pi P(k) l_k}{N+1})$, where the sum extends over the permutation group. We point out that, at each crossing point, the ground state jumps discontinuously in the spin Hilbert space from one symmetric subspace to another, orthogonal to the previous one.

4.4.2. *Finite size effects*

We now study ground state correlation, entanglement and fidelity at finite size. Due to the symmetry of the overall state, the one-spin reduced density matrix is purely diagonal, $\rho_l = \frac{1}{2}\text{diag}(1 + \langle\sigma_l^z\rangle, 1 - \langle\sigma_l^z\rangle)$; while for two spins at sites (l, m), one has $\rho_{lm} = a_+ |\uparrow\uparrow\rangle\langle\uparrow\uparrow| + a_- |\downarrow\downarrow\rangle\langle\downarrow\downarrow| + b_+ |\uparrow\downarrow\rangle\langle\uparrow\downarrow| + b_- |\downarrow\uparrow\rangle\langle\downarrow\uparrow| + e(|\uparrow\downarrow\rangle\langle\downarrow\uparrow| + |\downarrow\uparrow\rangle\langle\uparrow\downarrow|)$, with $a_\pm = \frac{1}{4}[1 \pm \langle\sigma_l^z\rangle \pm \langle\sigma_m^z\rangle + \langle\sigma_l^z \sigma_m^z\rangle]$, $b_\pm = \frac{1}{4}[1 \pm \langle\sigma_l^z\rangle \mp \langle\sigma_m^z\rangle - \langle\sigma_l^z \sigma_m^z\rangle]$ and $e = \frac{1}{2}\langle\sigma_l^x \sigma_m^x\rangle$. Here, the local magnetization and transverse correlation are given by $\langle\sigma_l^z\rangle = 1 - g_{l,l}$

and $\langle\sigma_l^z\sigma_m^z\rangle = (1-g_{l,l})(1-g_{m,m}) - g_{l,m}^2$, where

$$g_{l,m} = 2\sum_{r=1}^{k} S_l^r S_m^r = \frac{S_l^{k+1}S_m^k - S_l^k S_m^{k+1}}{2\left[\cos\left(\frac{\pi l}{N+1}\right) - \cos\left(\frac{\pi m}{N+1}\right)\right]}, \tag{4.69}$$

that depends on the field B through the index k.

The longitudinal correlation function $\langle\sigma_l^x\sigma_m^x\rangle$ is [149]:

$$\langle\sigma_l^x\sigma_m^x\rangle = \begin{vmatrix} G_{l,l+1} & G_{l,l+2} & \cdots & G_{l,m} \\ G_{l+1,l+1} & G_{l+1,l+2} & \cdots & G_{l+1,m} \\ \vdots & \vdots & \ddots & \\ G_{m-1,l+1} & G_{m-1,l+2} & \cdots & G_{m-1,m} \end{vmatrix} \tag{4.70}$$

where $G_{l,m} = \delta_{l,m} - g_{l,m}$. This determinant becomes of the Toeplitz type in the thermodynamic limit. For nearest neighbors, we simply get $\langle\sigma_l^x\sigma_{l+1}^x\rangle = G_{l,l+1}$.

With these density matrices at hand, we discuss the entanglement encoded in the state. Entanglement between a single spin and the rest of the chain can be measured by the one-tangle $\tau_l = 1 - \langle\sigma_l^z\rangle^2$ [150]. This quantity depends on the site for finite chain, and, as a function of B, it displays jumps at each crossing point B_k. Specifically: (i) τ_l equals one at zero field for every site; (ii) the jumps near $B = \pm 1$ become higher and higher moving from the end points towards the center of the chain, see Fig. 4.2(a). This feature of τ_l implies that the magnetic field effectively drives the system to a regime where the end-spins are more entangled than bulk ones.

To evaluate pairwise entanglement, we use the concurrence [150], $C_{l,m} = 2\max\{0, |e| - \sqrt{a_+a_-}\}$. Its behavior as a function of B for nearest neighboring spins is shown in Fig. 4.2(b), where jumps at crossing points are clearly visible. Moreover, for small values of B, the pairwise entanglement is bigger near the end points of the chain; while the reverse occurs at the borders of the critical region (see the discussion below).

The QFT can be further analyzed by studying the quantum fidelity of ground states with slightly different magnetic fields, B and $\tilde{B} = B + \delta B$. Specifically, we consider the partial state fidelity of reduced density matrix $\rho_a(B) = \mathrm{Tr}_b\rho(B)$ when the system is partitioned as $a + b$: $F_a(B,\tilde{B}) = \mathrm{Tr}\sqrt{\sqrt{\rho_a(B)}\rho_a(\tilde{B})\sqrt{\rho_a(B)}}$. F characterizes the degree of change of the state as the field is varied. For our purposes, it is sufficient to consider

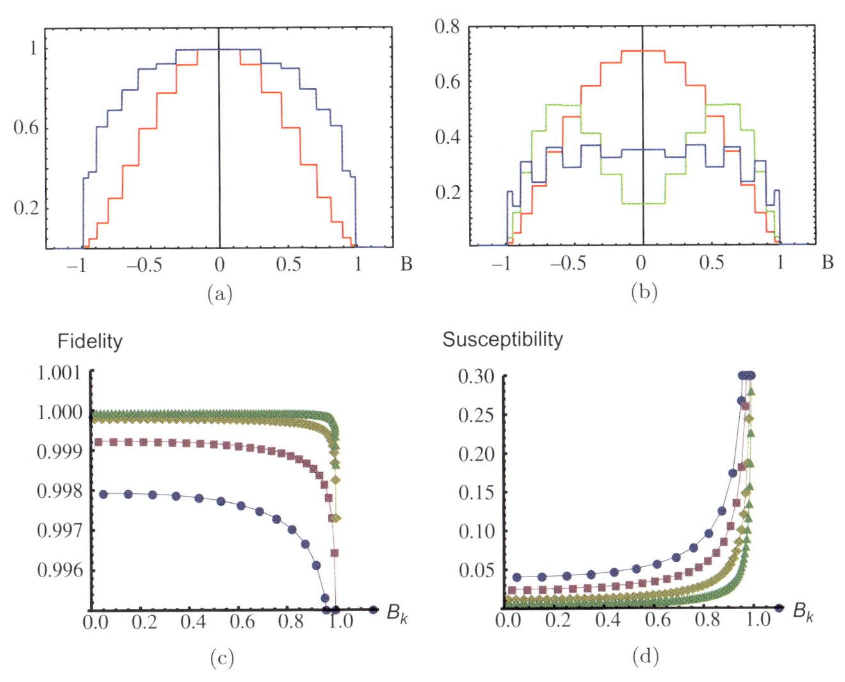

Fig. 4.2. (a) One-tangle for spin at site l as a function of the magnetic field for $l = 1$ (red line) and $l = 9$ (blue). (b) Comparison between nearest neighbor concurrence at the beginning and at the center of the chain. In the plot, concurrence is displayed between spins 1 and 2 (red line), 2 and 3 (green line), and 9 and 10 (blue). They are plotted when $N = 19$. (c) Partial state fidelities and (d) partial state fidelity susceptibilities as a function of external magnetic field $B_k = \cos(\frac{\pi k}{N+1})$ for different N. They are for the discrete values of odd integer k with large N, $N = 30$ (blue), $N = 50$ (red), $N = 100$ (yellow) and $N = 200$ (green) at $l = (N+1)/2$. The fidelity approaches to 1 as N increases.

the subsystem a to consist of just the lth spin [151]. In this case, F_l is a function of B via the local magnetization $\langle\sigma_z^l\rangle$. We find that, within the critical region, F_l is unit everywhere, except for a series of discrete and sharp drops at the crossing points $B = B_k$, odd k when $l = (N+1)/2$. For large system sizes, when the crossings become dense in the interval $|B| \leq 1$, we can perform a coarse-graining and evaluate F at these points. The result of this procedure is shown in Fig. 4.2(c), where one can see that only the drop at $B = 1$ remains, while all of the intermediate ones are smeared out. Interestingly enough, this behavior occurs only for bulk spins. If, instead, one of the end spins is singled out, the coarse grained fidelity stays flat.

An even more direct evidence that for large N the state of the bulk spins changes essentially in a continuous way except for the critical point $B = 1$, is obtained by looking at the fidelity susceptibility [151]: $\chi_F^l = \lim_{\delta B \to 0} -2 \ln F_l/(\delta B)^2$, which is plotted (for a bulk spin) in Fig. 4.2(d). This behavior appears related to the fact that, in the thermodynamic limit, the block entropy shows an essential singularity at $B = 1$ [152].

4.4.3. *Infinite spin chain*

For $N \to \infty$, the intervals $B_{k+1} < B < B_k$ become infinitesimally small and $\omega \equiv (\pi k)/(N+1)$ becomes continuous, $\omega \approx \arccos(B)$, so that the sum in (4.67) gives an energy per spin

$$\lim_{N\to\infty} \frac{\epsilon_g(B)}{N} = \frac{2}{\pi}[B(\arccos(B) - 1) - \sqrt{1 - B^2}], \tag{4.71}$$

which is analytic everywhere within the critical region, except for $B = \pm 1$. From the finite size analysis, however, we know that such region consists of dense set of crossings points (see Fig. 4.3(a)) and therefore can be considered as a line of continuous QFT, with the ground state driven by B through various symmetric spaces with $k \approx (N + 1)(\arccos B)/\pi$ flipped spins. The correlation functions and the entanglement reflect this behavior of the ground state.

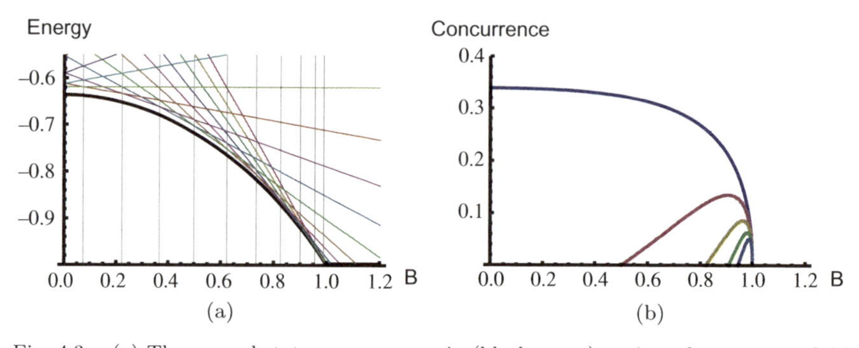

Fig. 4.3. (a) The ground state energy per spin (black curve) against the magnetic field B in the thermodynamic limit. The energy is plotted together with finite spin chain, $N = 20$. The intersections of the eigenvalues for the finite chain occur at the crossing points $B_k = \cos[k\pi/(N + 1)]$ (vertical lines) and the continuous energy level crossings in the infinite chain is appeared at the region of $0 < B < 1$. (b) Thermodynamic limit concurrence between two spins at distance $r = 1$ (blue), $r = 2$ (red), $r = 3$ (yellow), $r = 4$ (green) and $r = 5$ (cobalt). Entanglement between any two spins decreases as the distance between the spins increases.

The correlation function behave differently for spins close to the chain boundaries and for bulk spins. Setting the distance $r = m - l$, the limit $N \to \infty$ leads to different behaviors of $\langle \sigma_l^\alpha \sigma_{l+r}^\alpha \rangle$ depending on whether $l \gg r$ (bulk spins) or $l \ll r$ (end point spins), [153]. In particular, for $B = 0$ ($\omega = \pi/2$), $\langle \sigma_l^z \sigma_{l+r}^z \rangle = (1/r^2)(l+r)^2/(l+r/2)^2$, with a long distance behavior independent of boundary conditions. In contrast, longitudinal correlations are sensitive to the boundary

$$\langle \sigma_l^x \sigma_{l+r}^x \rangle = \begin{cases} \sqrt{2}\tilde{A}^2 r^{-1/2}, & l \gg r, \\ 4K(\frac{l+1}{2})(\sqrt{2})\tilde{A}^2 r^{-3/4}, & l \ll r, \end{cases} \tag{4.72}$$

where $\tilde{A} = 0.6450025$ and $K(x)$ is a function given in [153].

In the thermodynamic limit we have $g_{l,l+r} \simeq 2/\pi[\sin \omega r/r - \sin \omega(2l + r)/(2l+r)]$, so that, for example, for bulk spins at $B = 0$, we find $\langle \sigma_l^z \sigma_{l+r}^z \rangle = [1 - (-1)^r]\frac{(-2)}{\pi^2 r^2}$ and $\langle \sigma_l^x \sigma_{l+r}^x \rangle = (\frac{2}{\pi})^r \prod_{j=1}^{[r/2]-1} (\frac{4j^2}{4j^2-1})^{r-2j}$ where $[x]$ is the closest larger integer of x, which agrees with [154].

The concurrence $C_{l,l+r}$ can be derived from the above correlation functions. We find that $\lim_{l\to\infty} C_{l,l+r}$ disappears for $r \geq 2$. Only two nearest spins are entangled, with $\lim_{l\to\infty} C_{l,l+1} = 0.339$ at $B = 0$ [155]. Bulk concurrence $\lim_{l\to\infty} C_{l,l+r}$ between two spins at distance $1 \leq r \leq 5$ is plotted for various values of B in Fig. 4.3(b), where the decay of entanglement with the distance is also shown. For nearest neighboring spins, the concurrence goes $\sim 1/B$, disappearing at $B = 1$. Interestingly, as B is increased, bipartite entanglement starts to appear for every r and any two spins in the chain become entangled near $B = 1$, although the magnitudes become smaller and smaller with r. This is because $B = 1$ is the factorizing point for the XX model, with diverging entanglement range [156]. The one-tangle $\lim_{l\to\infty} \tau_l$ behaves in a similar way.

4.4.4. *Concluding summary*

In this section, we have analyzed the quasi-long-range order in the ground state of the XX model with transverse field in terms of quantum correlations and fidelity both for finite size and for an infinite chain. We have shown that the magnetic field drives the ground state of system through different Hilbert space sectors labeled by different eigenvalues of the total magnetization $\sum_i \sigma_i^z$.

We can also describe the critical phase exploiting a dual operator $\mu_n = \sum_{m<n} \sigma_m^x$. Once applied to a fully polarized state, it creates a

topological excitation of the system (a kink) [144, 157]. Indeed, any state with n spin flips can be viewed as suitable combination of kink–antikink pairs $\mu_n\mu_{n+1}$. When $B > 1$, all of the spins are aligned in the z direction and the state is separable; this is the state with no kinks. Near the critical point, with $B = 1 - \epsilon$, the ground state has only one flipped spin. Such a state is then a sea of condensed kink–antikink pairs (of infinite length in the thermodynamic limit). We have shown that in such a context all the spin pairs are entangled. By decreasing B, the size of the kink–antikink pairs decreases. At $B = 0$, a state is obtained with a single (degenerate) kink, with half of the spins pointing down and half pointing up. The state is "highly" symmetric and every spin is maximally entangled with the rest of the chain, but bipartite entanglement is present with the nearest neighbor only. This is due to the fact that the concurrence depends (and it is always smaller than) the longitudinal correlation function, which ultimately tends to zero because of the presence of the kinks. Indeed, even with quasi-long-range order (which arises because the spinwaves are massless), the long-range correlation function decays since the kinks are heavy.

At finite size the phenomenon of switching among states with a different number of kinks is witnessed by the sequence of jumps in the entanglement. Kinks, however, are essentially bulk excitations and the picture above is modified near the boundaries. Indeed, surface spins share entanglement differently from the bulk ones: for small magnetic fields, every spin is highly entangled with the rest, but the end-spins participate essentially to bi-partite entanglement, while the bulk ones are more involved in multi-partite correlations. On the other hand, for values of B near $|B| = 1$, surface spins are more entangled than those at the center, but they share multipartite rather than pairwise quantum correlations. Due to the changing number of kinks (or of flipped spins) the critical region is an instability line, with a marked anomaly at $B = \pm 1$ where the singular behavior persists even for $N \to \infty$, as evidenced by the partial state fidelity.

Summarizing, here we have identified the relevance of the kinks excitations in the ground state of the XX model from the perspective of phase transitions and entanglement. Preliminary analysis have shown that the formalism we have developed in the present paper open the way for the understanding of the role of cluster type states close to quantum phase transitions [158], especially for quantum information purposes.

4.5. Witnessing Macroscopic Entanglement: In a Staggered Magnetic Field

Here, we discuss how the extended example of the exactly soluble model can be treated in order to identify the effect of anisotropic magnetic field. In particular, solid-state quantum computation has become a topic of much research and several proposals for physical implementation have been investigated.

The Heisenberg interaction is the model used in many physical applications, for example, trapped ion, quantum dots [104] and cavity QED [105]. It has also been shown that the Heisenberg interaction can be used to implement any circuit required by a quantum computer [106]. Therefore, entanglement in 1D spin chains has been the subject of much interest. This entanglement has been studied both in the case of a finite spin chain [107, 108] and in the thermodynamic limit [109] where the length of the spin chain becomes infinite.

Macroscopic entanglement is a more recent concept. It demonstrates that non-local correlations persist even in the thermodynamic limit. This type of entanglement can be detected by measuring macroscopic quantities such as internal energy and magnetic susceptibility [114] as it has been proven that such quantities can be used as entanglement witnesses. It has been shown experimentally [49, 53], that the behavior of observable macroscopic quantities such as magnetic susceptibility depends, most significantly at low temperatures, on entanglement. This demonstrates that entanglement is vital in the explanation of how macroscopic materials behave. Macroscopic entanglement in a Heisenberg spin chain has been studied previously [109] only for a uniform magnetic field. The Hamiltonian of this chain is used to construct an entanglement witness [44, 110, 111] which shows that entanglement disappears for high uniform magnetic field just as it does for high temperature.

4.5.1. *Model description of the spins in inhomogeneous field*

In solid-state systems, there exists a possibility that an inhomogeneous Zeeman coupling could induce a non-uniform magnetic field. Moreover, an experimental system is likely to contain magnetic impurities. Copper Benzoate [112, 113] is a practical example of a system in a non-uniform magnetic field. In the case, the alternating field is in a direction perpendicular to the uniform field. Alternatively, such impurities could be introduced artificially. The possibility that such a field could affect entanglement, whether to reduce or increase it, is an important subject to investigate. In reality, systems have a finite temperature so the thermal case must be considered. Hence, here, we discuss the effect of a site-dependent magnetic field on thermal macroscopic entanglement in a 1D infinite spin-$\frac{1}{2}$ chain. We also consider the zero temperature case. Interestingly, we show that an alternating magnetic field can compensate for the effect of a uniform magnetic field at $T = 0$.

The Hamiltonian considered is

$$H = -\sum_{l} \left[\frac{J}{2} \left(\sigma_l^x \otimes \sigma_{l+1}^x + \sigma_l^y \otimes \sigma_{l+1}^y \right) + B_l \sigma_l^z \right], \tag{4.73}$$

where J is the coupling strength between sites, and $B_l = B + e^{-i\pi l} b$ is the site-dependent magnetic field. Although such a field is not likely to occur naturally, our work allows us to investigate how a non-uniform magnetic field affects entanglement in a model which is analytically solvable. A similar Hamiltonian with cyclic boundary conditions has previously been diagonalized [115] using a method first set out by Katsura [116, 117]. In our discussions of a finite N-spin chain, we consider the case of open boundary conditions with N even. In fact, these constraints are no longer relevant in the thermodynamic limit and hence our conclusion is unchanged if, for example, the cyclic model is used.

To identify entanglement in this system, we use an entanglement witness, i.e. an operator whose expectation value is bounded for any separable state. The power of our witness is such that we can identify the existence of entanglement even for a thermal system which is a mixed state in general. Alternatively, single-site entropy can be taken as evidence of entanglement when $T = 0$ since the total system is in a pure state. The purity of the single-site density matrix shows that the entanglement witness is not optimal at absolute zero. Thus we find that although the alternating magnetic field, b, acts in general to suppress entanglement similarly to B and T, at zero temperature, increasing b allows the system to be entangled for arbitrarily large B. Hence the effect of the staggered field at $T = 0$ is to increase both the amount and the region of entanglement.

4.5.2. *Entanglement witness and partition function*

As it is discussed, a system is said to be entangled when its density matrix cannot be written as a convex sum of product states. For a pure state, dividing the system into two subsystems A and B allows the von Neumann entropy to be used as a measure of entanglement. If we trace section B out of the density matrix to find ρ_A, the von Neumann entropy, $S(\rho_A) = -\mathrm{Tr}(\rho_A \log_2 \rho_A)$, can be calculated. $S(\rho_A) = 0$ corresponds to a separable state while when $S(\rho_A) = 1$, the system is maximally entangled. In the case of a mixed state, there is no unique measure of entanglement for a multipartite system. However, we can construct an entanglement witness.

An entanglement witness [118] is an operator whose expectation value for any separable state is bounded by a value corresponding to a hyperplane in the space of density matrices. An entanglement witness is only a sufficient condition for the existence of entanglement. Hence failure of the witness to detect entanglement does not necessarily mean the system is separable. Though witnesses simply detect rather than give a measure of entanglement, they have significant advantages over other methods. For example, they naturally incorporate temperature and many witnesses, such as magnetic susceptibility, can be experimentally measured [114].

Many thermodynamic variables can be derived from the Helmholtz free energy, $F = -T \ln Z$, where Z is the partition function. As $\partial F/\partial X = \langle \frac{\partial H}{\partial X} \rangle$, we see that when $X = B$, we obtain the magnetization $M = \sum_l \langle \sigma_l^z \rangle = -\partial F/\partial B$. In particular, we define the entanglement witness

$$W = \frac{2}{\beta N}\frac{\partial \ln Z}{\partial J} = \frac{1}{N}\sum_l (\langle \sigma_l^x \sigma_{l+1}^x \rangle + \langle \sigma_l^y \sigma_{l+1}^y \rangle), \tag{4.74}$$

where $\beta = 1/T$. Our witness, W, identifies a larger entangled region than witnesses found previously [109]. In a separable state, it satisfies the bound $|W| \leq 1$, which can be shown as follows. With $\rho = \sum_i p_i \rho_1^i \otimes \rho_2^i \otimes \cdots \otimes \rho_N^i$, we have $|\langle \sigma_l^x \sigma_{l+1}^x \rangle + \langle \sigma_l^y \sigma_{l+1}^y \rangle| = |\ \langle \sigma_l^x \rangle \langle \sigma_{l+1}^x \rangle + \langle \sigma_l^y \rangle \langle \sigma_{l+1}^y \rangle| \leq \sqrt{\langle \sigma_l^x \rangle^2 + \langle \sigma_l^y \rangle^2}\sqrt{\langle \sigma_{l+1}^x \rangle^2 + \langle \sigma_{l+1}^y \rangle^2} \leq 1$ for any l. The upper bound for the inequality is found by using the Cauchy–Schwarz inequality and the condition that for any state, $\langle \sigma_l^x \rangle^2 + \langle \sigma_l^y \rangle^2 + \langle \sigma_l^z \rangle^2 \leq 1$. Thus, any state that violates the inequality $|W| \leq 1$ is entangled.

In order to find the partition function of the system, we must diagonalize the Hamiltonian equation (4.73) and find its energy eigenvalues. The open ended Hamiltonian can be exactly diagonalized in a standard way via several steps: a Jordan–Wigner transformation, a Fourier transformation and finally a Bogoliubov transformation. The Jordan–Wigner transformation,

$$a_l = \prod_{m=1}^{l-1} \sigma_m^z \otimes \frac{(\sigma_l^x + i\sigma_l^y)}{2}, \tag{4.75}$$

maps the Pauli spin operators into fermionic annihilation and creation operators a_l and $a_l^\dagger$. These satisfy the anticommutation relations $\{a_l, a_k\} = 0$ and $\{a_l, a_k^\dagger\} = \delta_{l,k}$. Preserving the anticommutation relations, the operators can now be transformed unitarily using a Fourier transformation, $a_l = \sqrt{\frac{2}{N+1}} \sum_{k=1}^{N} d_k \sin(\frac{\pi k l}{N+1})$ and by a Bogoliubov transformation,

$$d_k = \alpha_k \cos\theta_k + \beta_k \sin\theta_k, \tag{4.76}$$

$$d_{N+1-k} = \beta_k \cos\theta_k - \alpha_k \sin\theta_k.$$

Setting $\tan 2\theta_k = b/[J\cos(\pi k/(N+1))]$ eliminates the off-diagonal terms leaving the Hamiltonian in diagonal form

$$H = \sum_{k=1}^{N/2} (\lambda_k^+ \alpha_k^\dagger \alpha_k + \lambda_k^- \beta_k^\dagger \beta_k - 2B\mathbf{1}), \tag{4.77}$$

where $\lambda_k^\pm = 2B \pm 2\sqrt{J^2\cos^2(\pi k/(N+1)) + b^2}$. The operators α_k and β_k satisfy the anticommutation relations $\{\alpha_k, \alpha_l^\dagger\} = \{\beta_k, \beta_l^\dagger\} = \delta_{k,l}$ and $\{\alpha_k, \beta_l\} = \{\alpha_k, \beta_l^\dagger\} = 0$. Using the eigenvalues of the Hamiltonian, we find that the partition function can be written $Z = \prod_{k=1}^{N/2} 2\cosh(\beta\lambda_k^+/2) 2\cosh(\beta\lambda_k^-/2)$. In the thermodynamic limit, $N \to \infty$, we can treat $\omega = \pi k/N$ as a continuous variable, to find

$$\ln Z = \frac{N}{\pi} \int_0^{\pi/2} d\omega \ln\left[4\cosh\left(\frac{\beta\lambda_\omega^+}{2}\right)\cosh\left(\frac{\beta\lambda_\omega^-}{2}\right)\right], \tag{4.78}$$

where $\lambda_\omega^\pm = 2B \pm 2\sqrt{J^2\cos^2\omega + b^2}$. We can now use Eq. (4.74) to calculate the entanglement witness for our system.

The region of entanglement detected by our witness has been plotted in Fig. 4.4. The figure shows the region of uniform magnetic field, B, temperature, T, and alternating magnetic field, b, within which we always find entanglement. At fixed values of b, the entangled region of the $T-B$ plane shrinks as b increases until a critical value is reached above which entanglement is no longer detected. Consider now a plane perpendicular to the T-axis. At zero temperature we see that until the critical value $b_c = 0.56$, increasing b has no effect on the value that the uniform magnetic field can take with the system remaining in an entangled state. Above b_c, our witness detects no entanglement. However, we later show that this witness is not optimal at zero temperature.

Our witness shows entanglement behaving as we would expect physically. A high temperature causes the system to become mixed. Thus

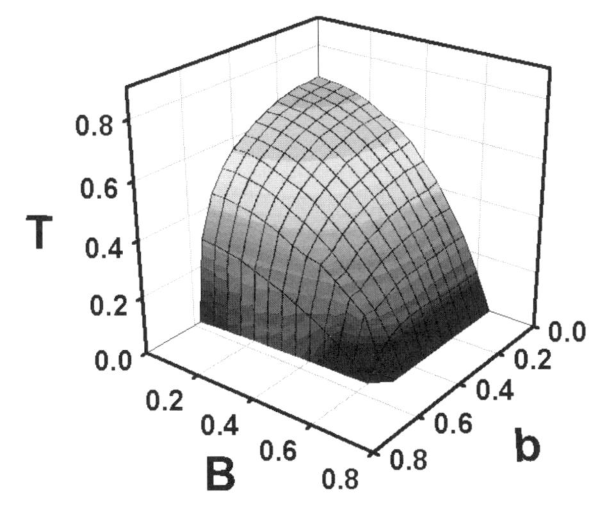

Fig. 4.4. The surface $|W| = 1$ in the space of magnetic field, B, temperature, T, and alternating magnetic field, b. The region between the surface and the axes is entangled.

no quantum correlations can survive and the system is separable. Moreover, the effect of the magnetic fields is understandable as local operations which enhance classical correlations. When the uniform magnetic field becomes large, spins tend to line up in a direction parallel to that field. This clearly decreases quantum correlations in the system. Using the same reasoning, if the alternating magnetic field is large, the spins tend to antialign which is also a product state. Hence, as Fig. 4.4 shows, all of these parameters cause the system to become separable if they are large enough. Interestingly, we find a counter example of this expected behavior and identify a region where the system is entangled even in a large magnetic field. We discuss this in the following section.

4.5.3. *Single-site entropy*

When $T = 0$, the total system is in a pure state, i.e. the ground state. Thus, if the density matrix of a single spin is in a mixed state, the particle must be entangled with the rest of the spins. This can be quantified using the entropy of a single spin. The lth spin density matrix, $\rho_l = \frac{1}{2}\sum_{i\in\{1,x,y,z\}} \sigma_l^i \langle\sigma_l^i\rangle$, can be obtained from the total density matrix $\rho = e^{-\beta\hat{H}}/Z$. Moreover, we find that the single-site density matrix is readily diagonalized since $\langle\sigma_l^x\rangle = \langle\sigma_l^y\rangle = 0$. This follows from the fact that σ_l^x and σ_l^y are linear combinations of the fermion operators α_k, $\alpha_k^\dagger$, β_k, and $\beta_k^\dagger$, all of which have zero expectation values.

In the thermodynamic limit $N \to \infty$, the system is translationally invariant for all odd sites and for all even sites. Hence the single-site magnetization $\langle \sigma_l^z \rangle$ can be obtained from the total magnetization M and the total staggered magnetization M_s. In the limit of zero temperature, these are given by

$$M = \sum_l \langle \sigma_l^z \rangle = \frac{1}{\beta}\frac{\partial}{\partial B} \ln Z = N\left(1 - \frac{2\Omega}{\pi}\right), \tag{4.79}$$

$$M_\mathrm{s} = \sum_l (-1)^l \langle \sigma_l^z \rangle = \frac{1}{\beta}\frac{\partial}{\partial b} \ln Z = \frac{N}{\pi}\int_0^\Omega d\omega f(\omega),$$

where $f(\omega) = 2b/\sqrt{J^2\cos^2\omega + b^2}$ and

$$\Omega = \begin{cases} \pi/2 & \text{for } B < b, \\ \cos^{-1}(\sqrt{B^2 - b^2}/J) & \text{for } b \le B \le \sqrt{J^2 + b^2}, \\ 0 & \text{for } B > \sqrt{J^2 + b^2}. \end{cases}$$

To obtain these results, we have used $\lim_{\beta\to\infty} \tanh(\beta x) = x/|x|$. By virtue of translational invariance, $\langle \sigma_l^z \rangle$ is the same for all even sites and for all odd sites. Thus defining $\langle \sigma_l^z \rangle = \mathrm{Tr}(\sigma_l^z \rho) = 2\eta_l - 1$, such that the single-site density matrix is $\rho_l = \mathrm{diag}(\eta_l, 1 - \eta_l)$, we have

$$\eta_l = \frac{1}{2}\left\{1 + \frac{1}{N}\left[M + (-1)^l M_\mathrm{s}\right]\right\}. \tag{4.80}$$

From this, we can obtain the entropy of the lth spin, $S = -\eta_l \log_2 \eta_l - (1 - \eta_l)\log_2(1 - \eta_l)$.

The single-site entropy for odd and even sites is plotted in Fig. 4.5 with $J = 1$. In both cases, we find entanglement for various values of B and b even when the magnetic fields are large. The entropy, and therefore entanglement between each spin and the remainder of chain, is non-zero everywhere except when $B > \sqrt{J^2 + b^2}$. Hence entanglement exists when the coupling strength between nearest neighbor spins, J, is more than $\sqrt{B^2 - b^2}$. We note that this corresponds to when the square of the interaction strength is greater than the product of the total magnetic field on two adjacent sites. In addition, we observe from Fig. 4.5 that the maximum single-site entropy occurs when both magnetic fields B and b are zero. Introducing the magnetic fields reduces the amount of entanglement in the system

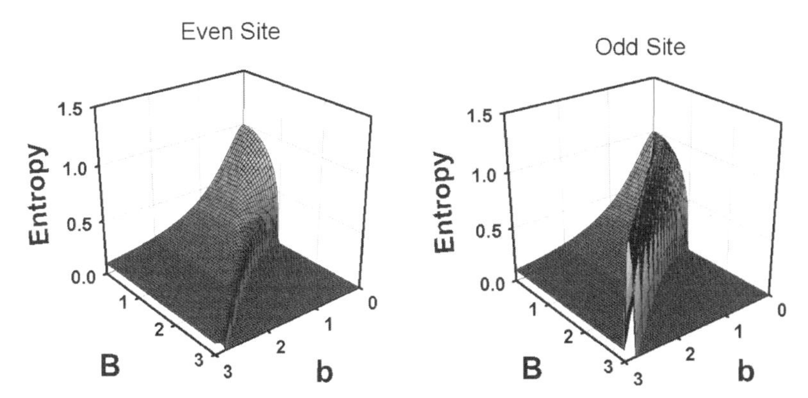

Fig. 4.5. We plot the entropy for an even site spin and an odd site spin respectively when $T = 0$ and $J = 1$. Excluding the region $B > \sqrt{1+b^2}$, the single-site entropy is non-zero and each spin is entangled with the rest of the system. The odd site entropy shows a maximum when $B = b + \varepsilon$.

except along the peak in the odd site entropy. We find from the maximum entropy, $S_l = 1$, that the maximum entanglement occurs in the region $b \leq B \leq \sqrt{J^2 + b^2}$ when $\Omega - \pi/2 = (-1)^l \int_0^\Omega b/\sqrt{J^2 \cos^2 \omega + b^2} d\omega$ is satisfied. This corresponds to when the magnetization, M, is equal to the staggered magnetization, M_s. For even sites, there is only one solution to this at $B = b = 0$. For odd sites however, this is satisfied for any finite uniform magnetic field B at $B = b + \varepsilon$ where ε is a positive value. The solutions in Fig. 4.6 correspond to the peak in the odd site entropy and so occur within the range $b \leq B \leq \sqrt{J^2 + b^2}$. In general, ε becomes larger as J increases as shown in Fig. 4.6, and becomes smaller as B and b increase. We note that the peak does not occur at $B = b$. As B and/or b tend to infinity, the amount of entanglement in the system tends to zero. Further, as the magnetic fields increase, the curves in Fig. 4.6 tend to the $B = b$ line. At $B = b$, the system is no longer maximally entangled. At this point, we find from the Hamiltonian that odd sites have zero magnetic field, and even sites have field strength $2B$. Hence at the peak, odd site spins have a small magnetic field, ε, while in comparison, even site spins have a large field, $2B - \varepsilon$. The implications of this peak are that even in the limit of large (though not infinite) uniform field B, an odd site can be maximally entangled with the rest of the system if an appropriate alternating magnetic field, b is introduced. In practice, fine-tuning b to achieve maximal entanglement may be difficult. However, for any B, sufficiently increasing b (when $b > \sqrt{B^2 - J^2}$) will create entanglement

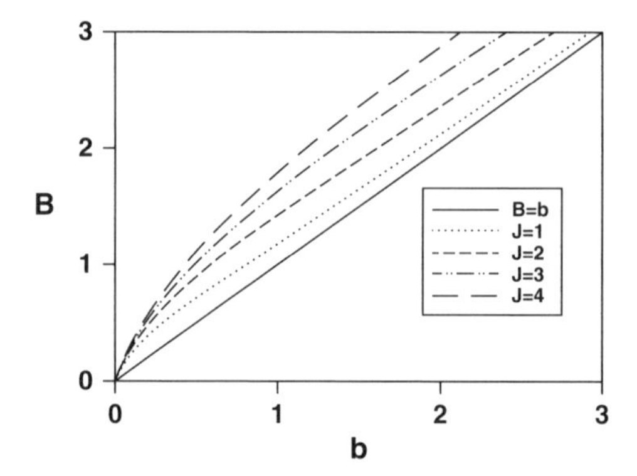

Fig. 4.6. Plot of the maximum entropy $S_l = 1$ for odd site spins with different values of J. As the value of J increases, this maximum entropy moves further away from the line $B = b$.

in the chain. Hence a staggered magnetic field can enhance the amount of entanglement present in the system.

Although the single-site entanglement relates only to the zero temperature case, changing the temperature by a small amount should not change its behavior. Hence for very low temperatures, we see that the entanglement witness is not optimal.

4.5.4. *Section remarks*

Our best estimate for finite temperature entanglement is the witness which shows b reduces the region of entanglement in the chain. That is, a staggered magnetic field reduces the entangled region. If this behavior is true even for an optimal finite temperature witness, these results have consequences for larger scale quantum computation in solid-state systems. As inhomogeneities in the magnetic field exist naturally in the Zeeman coupling between atoms, the region of entanglement is naturally decreased compared to when $b = 0$. Quantum computation relies on entanglement so as introducing b reduces both the temperature and uniform magnetic field at which entanglement persists, our result shows it may be more difficult than previously thought to construct useful quantum computers using 1D systems.

Our entanglement witness is invaluable as by applying it to our system, it allows us to see how temperature affects the entanglement in the spin chain. However, the witness does not tell us how the entanglement actually behaves in the presence of B, T or b as it does not detect all entanglement in the system. Conversely, the single-site entropy shows us exactly how the entanglement is affected by the magnetic fields at zero temperature, although it is unknown how to extend this entropy to a finite temperature. Using this entropy, we have shown that in the thermodynamic limit, a staggered magnetic field enhances both the region and the amount of entanglement in our spin chain. Hence, both the witness and the entropy are essential in characterizing the entanglement in the system. Further, the region of entanglement identified by the witness is consistent with that of the single-site entropy. If the behavior of the entanglement as shown by the entropy persists at higher temperatures, we may be able to counteract any Zeeman coupling by applying an appropriate magnetic field, hence maximizing entanglement for odd sites. This will be an interesting topic for future research.

4.6. Matrix Product State as a New Approach

Description of many-body system has been known as an intractable problem in quantum physics. Although simple 1D many-body models can be exactly solved by the numerous efforts, it is still not enough to overcome the tractability of various models in many-body quantum systems. Since the 1990s, more progress in the many-body simulation has been made since tensor network theories have drastically expanded to solve a variety of the many-body problems efficiently. Matrix product state (MPS) method for 1D systems and projected entangled pair state (PEPS) method for 2D systems are, especially, widely used to describe the quantum states and can be applied to calculate the various physical quantities. They are to evaluate the ground energy, the two-point correlation function and the bipartite entanglement measures, which gives the insight of the properties of the many-body systems. The analytic approaches are originated from the density matrix renormalization group (DMRG) method in condensed matter physics [163]. Since the method, there were various tries to represent the quantum states from DMRG as a matrix of tensor product states. Within the framework of quantum information, MPS method is proved to be more efficient than DMRG method for the periodic boundary condition.

4.6.1. *Basic concept of MPS*

MPS representation intuitively shows the structure of the entanglement between two subsystems. In this method, the exponential order of the quantum system $O(\exp(N))$ can be reduced to the linear order $O(N)$, which results in the efficient simulation of the many-body quantum system. By using the MPS representation and the variational method, it is possible to construct the algorithms for searching the class of matrices pertaining to the many-body ground state. In this section, we are going to discuss the expression of the matrix product state for a given many-body state briefly.

Let us consider the simplest case: a pure state for two particles A and B in d-dimensional Hilbert space such that

$$|\psi\rangle = \sum_{i,j=1}^{d} c_{i,j}|i\rangle_A|j\rangle_B. \tag{4.81}$$

When the coefficient $c_{i,j}$ is considered as the matrix element, the pure state $|\psi\rangle$ can be expressed by using the singular value decomposition [52]

$$|\psi\rangle = \sum_{i,j=1}^{d}\sum_{\alpha=1}^{\chi} c_{i,j}|i\rangle_A|j\rangle_B, \tag{4.82}$$

which is nothing but the matrix product state for two particles. As finding the matrix product form of bipartite pure states by using the single value decomposition, a pure state for three subsystems also can be represented through the consecutive singular value decomposition. Suppose that there are three subsystems A, B, and C in d-dimensional Hilbert space. Then, a three-body quantum state can be written as the matrix product form

$$|\psi\rangle = \sum_{i,j,k=0}^{d-1} c_{ijk}|i\rangle_A|j\rangle_B|k\rangle_C \tag{4.83}$$

$$= \sum_{i,j,k=0}^{d-1}\sum_{\alpha_1,\alpha_2=0}^{\chi-1} U_{i\alpha_1}\Lambda_{\alpha_1\alpha_1}V_{\alpha_1 j\alpha_2}\Lambda'_{\alpha_2\alpha_2}W_{\alpha_2 k}|i\rangle_A|j\rangle_B|k\rangle_C, \tag{4.84}$$

where the coefficients c_{ijk} are regarded as the entries of a matrix. If the coefficients of a quantum pure state are given, the matrices in Eq. (4.84) can be found through the singular value decomposition twice, called the *Vidal's form* [159, 160].

If there are linearly aligned N particles with orthonormal bases in d-dimensional local Hilbert space, then a pure quantum state $|\psi\rangle$ can be written as

$$|\psi\rangle = \sum_{s_1,\ldots,s_N} c_{s_1,s_2,s_3\ldots,s_N} |s_1\rangle \otimes |s_2\rangle \otimes |s_3\rangle \otimes \cdots \otimes |s_N\rangle \tag{4.85}$$

$$= \sum_{\{s_i\}} \sum_{\{\alpha_i\}}^{D} \Gamma^{[1]s_1}_{\alpha_1} \Lambda^{[1]}_{\alpha 1 \alpha_1} \Gamma^{[2]s_2}_{\alpha_1\alpha_2} \Lambda^{[2]}_{\alpha_2\alpha_2} \Gamma^{[3]s_3}_{\alpha_2\alpha_3} \cdots \Gamma^{[N]s_N}_{\alpha_N} |s_1\rangle |s_2\rangle \cdots |s_N\rangle, \tag{4.86}$$

where $\{\Gamma^{[1]}, \Gamma^{[2]}, \ldots, \Gamma^{[N]}\}$ and $\{\Lambda^{[1]}, \ldots, \Lambda^{[N-1]}\}$ represent the decomposed matrices of the original tensor $c_{s_1,\ldots,s_N}$ with the physical index $s_i \in \{1, 2, \ldots, d\}$ and the auxiliary index $\alpha_i \in \{1, 2, \ldots, \chi_i\}$ (see Fig. 4.7) [160]. The matrix $\Lambda^{[i]}$ can be directly calculated from the singular value matrix of Ψ_{AB} whose element is $c_{(s_1\cdots s_i)(s_{i+1}\cdots s_N)}$. That is, the diagonal elements of $\Lambda^{[i]}$ correspond to the square root of the eigenvalues of the reduced density matrices $\rho^{[i]}_A (= \Psi_{AB}\Psi^\dagger_{AB})$ or $\rho^{[i]}_B (= \Psi^\dagger_{AB}\Psi_{AB})$ where A in the range of 1 to ith site and B with the rest of them are the subsystems of the entire system. The dimension of the row (or column) of the matrix $\Lambda^{[i]}$ is nothing but the Schmidt rank of the Ψ_{AB} denoted by χ_i. The value of χ_i increases

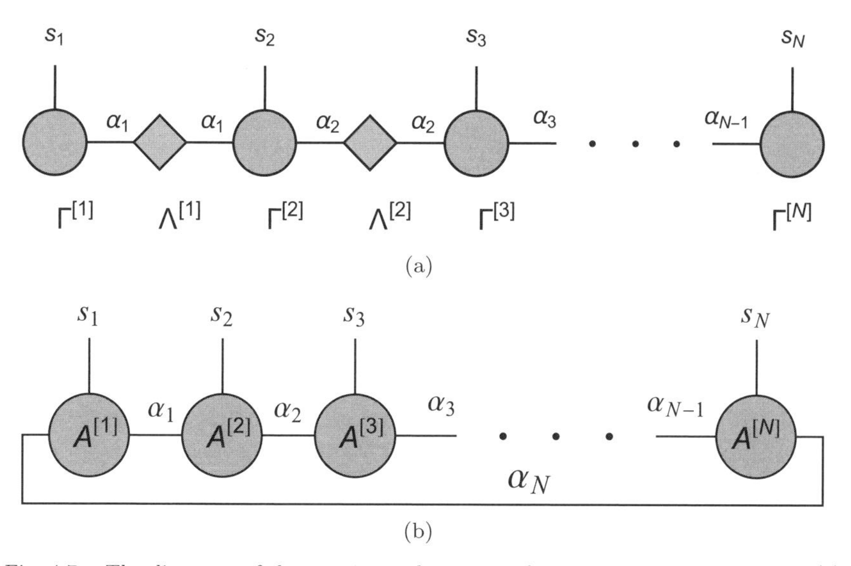

Fig. 4.7. The diagrams of the matrix product states for open boundary conditions (a) and periodic boundary conditions (b): circles and diamonds represent each matrices. Rods mean the indices of the matrix. If rod is connected between two matrices, then two matrices sharing the rod are multiplied with respect to the index corresponding to the rod.

exponentially from d to $d^{N/2}$ as i goes to $N/2$ under the open boundary condition, which has been a complex problem in the case of many-body quantum systems.

But it is fortunate that one can approximately approach this problem by fixing the exponentially growing dimension at a low number D, called the *bond dimension*. Fixing the value of D has an effect on the fact that the matrix $\Lambda^{[i]}$ diagonally consisting of the singular values in descending order is confined to the $D \times D$ matrix including up to the Dth singular value. All the auxiliary indices $\{\alpha_1, \alpha_2, \ldots, \alpha_N\}$ become in the range of 1 to D. The rest of them and the corresponding eigenvectors are discarded. The order of many-body quantum state are consequently reduced from $O(d^N)$ to $O(NdD^2)$.

Not only the Vidal's MPS form but also other MPS forms there are for the same pure state. For the open boundary condition, the *left-canonical* matrix product state is defined as the pure state consisting of left-normalized matrices, i.e.

$$|\psi\rangle = \sum_{s_1,\cdots,s_N} A^{s_1} A^{s_2} \cdots A^{s_N} |s_1, s_2, \ldots, s_N\rangle, \tag{4.87}$$

with the left-normalized matrices $\sum_{s_i} A^{s_i \dagger} A^{s_i} = \mathbb{1}$. In this manner, the definition of the *right-canonical* matrix product state is

$$|\psi\rangle = \sum_{s_1,\cdots,s_N} B^{s_1} B^{s_2} \cdots B^{s_N} |s_1, s_2, \ldots, s_N\rangle, \tag{4.88}$$

with right-normalized matrices $\sum_{s_i} B^{s_i} B^{s_i \dagger} = \mathbb{1}$. In connection with the Vidal's form, the left-normalized matrix A^{s_i} and the right-normalized matrices B^{s_i} have the relations with the matrices $\Gamma^{[i]s_i}$ and $\Lambda^{[i]}$ in the Eq. (4.86). That is,

$$A^{s_i} \equiv \Lambda^{[i-1]} \Gamma^{[i]s_i} \quad \text{and} \quad A^{s_1} \equiv \Gamma^{[1]s_1}, \tag{4.89}$$

$$B^{s_i} \equiv \Gamma^{[i]s_i} \Lambda^{[i]} \quad \text{and} \quad B^{s_N} \equiv \Gamma^{[N]s_N}. \tag{4.90}$$

4.6.2. *Boundary effects*

Up to now, it is mentioned that there are some kinds of MPS forms for the systems in the open boundary condition. It is possible to recognize that the quantum state for the periodic boundary systems can be written by using the MPS representation, which is different from one for the open boundary systems. In the periodic boundary condition, the left-canonical MPS becomes

$$|\psi\rangle = \sum_{s_1,\ldots,s_N} \mathrm{Tr}[A^{s_1} A^{s_2} \cdots A^{s_N}]\, |s_1, s_2, \ldots, s_N\rangle, \tag{4.91}$$

where the trace signifies the connection between the matrix A^{s_1} and A^{s_N} which are not actually the vectors but matrices under the periodic boundary conditions. In other words, the matrix product $A^{s_N} A^{s_1}$ exists. It is remarkable that the α_N rod connected between the first site and the last site is added in Fig. 4.7 showing the graphical notation for the periodic MPS in spatial 1D system. This description of the MPS for the periodic boundary condition overcomes the limitation of the DMRG method for the periodic boundary systems [161] and the spatial 2D systems [162].

With this description, one can describe the ground states for 1D and 2D many-body systems except the frustrated Hamiltonian, which are related to the various physical properties. In order to find the ground state for 1D systems, one can make use of the variational method, which needs the matrix product representation for operators. Specific procedure of computing the ground state is given in Appendices A.8 and A.9.

Appendix

A.1. The Continuum Limit

We start with the kinetic second quantized Hamiltonian:

$$H = \frac{\hbar^2}{2m} \int \Psi^\dagger(x) \nabla^2 \Psi(x) dx, \tag{A.1}$$

where $\Psi(x)$ represents the annihilation operator for a boson. Suppose that bosons are localized around some lattice points (e.g. positions of fixed atoms). Then $\Psi(x) \approx \sum_i \phi(x - x_i) b_i$ and $\Psi^\dagger(x) \approx \sum_i \phi^*(x - x_i) b_i^\dagger$, where $\phi^*(x)$ and $\phi(x)$ are the so-called Wannier functions. Substituting these into the Hamiltonian leads to

$$H = -J \sum_{ij} b_i^\dagger b_j + U \sum_i n_i(n_i - 1), \tag{A.2}$$

where $J = \hbar^2/2m \int \phi^*(x - x_i) \nabla^2 \phi(x - x_j) dx$, $U = 4\pi\hbar^2 s/m \int dx |\phi(x)|^4$ and s is the scattering length. If the ϕ function is well localized, then only the nearest neighbor coupling will be relevant, which is what we assumed throughout the paper. The on-site term is just a constant referring to the total number of atoms. If the density of atoms is low, then the n_i^2 term can be ignored and we obtain the XX Hamiltonian analyzed in the paper. The average energy per particle is, at low densities, given by $E \approx (h^2/2m)(\rho s)$. From here, the healing length, a, is defined by $a = 1/\sqrt{\rho s}$ so that the energy is $E \approx (h^2/2ma^2)$. This leads to the following entanglement witness that was mentioned in the text: if $\langle H \rangle < (h^2/2ma^2)$, then the system is entangled.[46] This leads to the critical temperature since $\langle H\rangle$ is a function of temperature. Approximately, $\langle H\rangle \approx h^2/2m\lambda_T^2$, where λ_T is the thermal wavelength, so the criterion for entanglement is that $\lambda_T > a$, which requires the thermal wavelength to be larger than the healing length.

A.2. Two-Dimensional XX Lattice

True signatures of superfluidity, such as quantized vortices, can really only exist in two dimensions and higher. We will see, however, that there will be no fundamental changes to our earlier 1D conclusions. The 2D XX Hamiltonian is given by

$$H = \sum_i \sum_j J(\sigma_{i,j}^x \sigma_{i+1,j}^x + \sigma_{i,j}^y \sigma_{i+1,j}^y) + J_\perp(\sigma_{i,j}^x \sigma_{i,j+1}^x + \sigma_{i,j}^y \sigma_{i,j+1}^y). \tag{A.3}$$

We do not know how to diagonalize this Hamiltonian exactly. However, we can do reasonably well by first applying generalized (2D) Jordan–Wigner transformation, and then a mean-field approximation on the phases. The Jordan–Wigner transformation in two dimensions is defined as follows [50]: $\sigma^- = e^{i\alpha_{i,j}} d_{i,j}$ and $\sigma^+ = e^{-i\alpha_{i,j}} d_{i,j}$. Applying the mean-field approximation on the phases α, we obtain the transformed Hamiltonian of the form [50]

$$\begin{aligned} H = \frac{1}{2} \sum_i \sum_j & J(-1)^{i+j}(d_{i,j}^\dagger d_{i+1,j} - d_{i,j} d_{i+1,j}^\dagger) \\ & + J_\perp(d_{i,j}^\dagger d_{i,j+1} - d_{i,j} d_{i,j+1}^\dagger), \end{aligned} \tag{A.4}$$

which, upon diagonalization by the 2D Fourier transform, gives [51]: $H = \sum_k \Lambda(k)(\eta_k^\dagger \eta_k - \frac{1}{2})$, where the eigenvalues are

$$\Lambda(k) = \sqrt{J_\perp^2 \cos^2 k_y + J^2 \sin^2 k_x}.$$

This now presents the (approximately) diagonalized 2D XX model. Let us now use the energy-based entanglement witness [42] to analyze this Hamiltonian. The energy can easily be calculated from the partition function and is given by $U = -\frac{1}{2}\int \frac{d^2k}{(2\pi)^2} \Lambda \tanh \frac{\beta\Lambda}{2}$. Applying our energy-based entanglement witness now reads: $|U_\rho| > (J + J_\perp)/2$ which implies that the state ρ is entangled. Let us take the low and high temperature limit of the energy formula to draw some conclusions on entanglement. At low T, $\tanh \to 1$ and the (absolute value of) energy is demonstrably larger than the separable bound. The ground state in two dimensions is therefore provably entangled. At high T, $\tanh x \to x$ and so $|U| \to \beta(J^2 + J_\perp)^2/8$, which implies entanglement for temperatures such that $kT < \frac{1}{8}\frac{J^2+J_\perp^2}{J+J_\perp}$, which is very similar to the 1D result.

A.3. Hamiltonian Transformation

In this section, we show how the Hamiltonian for the electron–phonon interactions can be transformed into the Hamiltonian describing electron–electron interactions. We verify that such an attraction between the electrons does exist by performing a canonical transformation of the Fröhlich Hamiltonian. We write

$$H = H_0 + H_{\mathrm{el-ph}} \tag{A.5}$$

and then look for a transformation of the form $H' = e^{-S}He^{s}$ that will eliminate $H_{\mathrm{el-ph}}$ to the first order. Upon expansion of the exponentials, we use Baker–Campbell–Hausdorff formula and have

$$H' = \left(1 - s + \frac{1}{2}s^2 + \cdots\right) H \left(1 + s + \frac{1}{2}s^2 + \cdots\right) \tag{A.6}$$

$$= H + [H, s] + \frac{1}{2}[[H, s], s] + \cdots . \tag{A.7}$$

We choose s in such a way that its commutator with H_0 cancels the term $H_{\mathrm{el-ph}}$ as $H_{\mathrm{el-ph}} + [H_0, s] = 0$. With the choice and the omission of the higher order than s^3, the transformed Hamiltonian becomes

$$H' = H_0 + \frac{1}{2}[H_{\mathrm{el-ph}}, s], \tag{A.8}$$

where $H_0 = \sum_k \epsilon_k c_k^\dagger c_k + \sum_q \hbar\omega_q a_q^\dagger a_q$, $H_{\mathrm{el-ph}} = \sum_{k,k'} M_{k,k'}(a_{-q}^\dagger + a_q)c_k^\dagger c_{k'}$ and we assume that

$$s = \sum_{k,k'} M_{k,k'}(Aa_{-q}^\dagger + Ba_q)c_k^\dagger c_k \tag{A.9}$$

with A and B coefficients to be determined. In order to satisfy $H_{\mathrm{el-ph}} + [H_0, s] = 0$, we must have

$$A = \frac{-1}{\epsilon_k - \epsilon_{k'} + \hbar\omega_{-q}}, \quad B = \frac{-1}{\epsilon_k - \epsilon_{k'} - \hbar\omega_q}. \tag{A.10}$$

Then, it leads to the transformed Hamiltonian,

$$H' = H_0 + \frac{1}{2}\left[\sum_{k,k'} M_{k,k'}(a_{-q}^\dagger + a_q)c_k^\dagger c_k, \times \sum_{k'',k'''} M_{k'',k'''}(Aa_{-q}^\dagger + Ba_q)c_{k''}^\dagger c_{k'''}\right], \tag{A.11}$$

where the second term generated many commutators. Using the Bosonic commutation relations of the phonon modes $[a_q^\dagger, a_{q'}] = \delta_{q,q'}$, one can collect the terms which survive in the commutator. With the fact that $M_{k,k'}$ is a function only of $k - k' = q$, we find

$$H' = H_0 + \sum_{k,k',q} |M_q|^2 \frac{\hbar\omega_q}{(\epsilon_k - \epsilon_{k-q})^2 - (\hbar\omega_q)^2} c_{k'+q}^\dagger c_{k-q}^\dagger c_k c_{k'}, \tag{A.12}$$

where the electron–electron interaction is decoupled from the phonon mode.

A.4. Evaluation of Self-Consistency Integral

By the replacement $\epsilon/k_B T$, the integral for the self-consistency can be

$$V\rho_f \int_0^{\hbar\omega_D/2k_BT} x^{-1}\tanh[x]dx = 1. \tag{A.13}$$

The integral in the left-hand side can be expanded by parts

$$\ln[x]\tanh[x]\big|_0^{\hbar\omega_D/2k_BT} - \int_0^{\hbar\omega_D/2k_BT} x^{-1}\cosh^{-2}[x]\ln[x]dx = \frac{1}{V\rho}. \tag{A.14}$$

For weak-coupling superconductors, we can replace $\tanh[\hbar\omega_D/2k_BT]$ by unity and extend the upper limit of the integral to infinity to find

$$\ln\left(\frac{\hbar\omega_D}{2k_BT}\right) - \int_0^{\infty} x^{-1}\cosh^{-2}[x]\ln[x]dx = \frac{1}{V\rho}. \tag{A.15}$$

The integral is evaluated and it gives the critical temperature

$$k_BT_c = 1.14\ \hbar\omega_D e^{-\frac{1}{\rho_f V}}. \tag{A.16}$$

A.5. Classical Origin of Current Equation

This macroscopic-order parameter has properties similar to a quantum mechanical wave function. If the order parameter varies locally as $\psi(r) = \psi_0 e^{iq\cdot r}$, then we can associate a momentum $2m_e v = \hbar q$ with each Cooper pair with the supercurrent density $j_s(r)$ given by $j_s(r) = \frac{n_s}{2}(-2e)\frac{\hbar q}{2m_e} = -|\psi(r)|^2 \frac{e}{m_e}\hbar q$. We must modify this definition of current density in the presence of an applied magnetic field B. The electromagnetic momentum p for a particle (Cooper pair) of mass $2m_e$ and charge $-2e$ are given as $p = 2m_e v - 2eA$, where A is the magnetic vector potential since $B = \nabla \times A$. The kinetic momentum is therefore given by $2m_e v = p + 2eA$ and the velocity v of a Cooper pair by

$$v = \frac{p}{2m_e} + \left(\frac{e}{m_e}\right) A. \tag{A.17}$$

To calculate the velocity v in a quantum mechanical analysis, we replace the total momentum p by the momentum operator $-i\hbar\nabla$. If we assume that this is also true for the superconducting order parameter, we might deduce that the superconducting current density $J(r)$ in the presence of a magnetic field B is given by

$$J(r) = -\frac{e}{m_e}\langle\psi(r)|[-i\hbar\nabla + 2eA]|\psi(r)\rangle. \tag{A.18}$$

In order for the local current density to be real, we must define the current as half the sum of the current density and its complex conjugate,

$$j(r) = [J(r) + J(r)^*]/2, \tag{A.19}$$

which becomes the electric current.

A.6. Proving London Equation

Here, we inspect the quantum mechanical version of London equation which states that the electric current is proportional to the electric density. From the electrodynamics, Hamiltonian of a system in an electric field is

$$H = \frac{1}{2m}\left(p - \frac{e}{c}A\right)^2 + V(r). \tag{A.20}$$

The Hamiltonian can be divided into two parts as $H = H_0 + H_I$. Here H_0 is the free electron part and H_I is the perturbation due to the applied field as

$$H_0 = \frac{1}{2m}\left(p - \frac{e}{c}A\right)^2 + V(r), \tag{A.21}$$

$$H_I = \frac{e}{2mc}\left(\frac{e}{c}A^2 - p\cdot A - A\cdot P\right) \tag{A.22}$$

when the relativistic effect is neglected. If we drop the term in A^2, then the perturbation term can be written in a second quantization form as

$$H_I = \frac{2}{2mc}\sum_{k,k'}\langle k|\left(-p\cdot A - A\cdot P\right)|k'\rangle c_k^\dagger c_{k'} \tag{A.23}$$

$$= \sum_{k,k'}\langle k|A(k+k')|k'\rangle c_k^\dagger c_{k'}, \tag{A.24}$$

where we omit the spin degree of freedom in the Fermi creation and annihilation operator. When the vector potential is of the form $A_p e^{ip\cdot r}$, we have

$$H_I = -\mu_B A_p \cdot \sum_k (2k - p) c_k^\dagger c_{k-p} \tag{A.25}$$

with $\mu_B = e\hbar/2mc$. The current density ev/V is found from the continuity equation

$$V^{-1}\nabla\cdot v = -\frac{\partial\rho}{\partial t}, \tag{A.26}$$

of which the Fourier transform is

$$ip\cdot v_q = -\frac{\partial\rho_p}{\partial t} = \frac{1}{i\hbar}[H,\rho_p], \tag{A.27}$$

where the density function is $\frac{1}{V}\sum_p c_{p+q}^\dagger c_{p+q'}$. With the multiplication of the electric charge e, we thus find

$$p\cdot j_p = -\frac{e}{\hbar}[H_0 + H_I, \rho_p]. \tag{A.28}$$

Using the Fermi anticommutation relation, the commutator of the zero-field Hamiltonian H_0 with ρ_p is able to be evaluated and its result is

$$j_p = \sum_k\left[\frac{e\hbar}{2mV}(2k-q)c_{k-q}^\dagger c_k - \frac{e^2 A_p}{mcV}c_k^\dagger c_k\right]. \tag{A.29}$$

The effect of the magnetic field is to perturb the wave function of the superconductor from its initial state $|\psi_0\rangle$ to the new state $|\psi\rangle$ that is given to the first-order perturbation in H_I by the usual prescription

$$|\psi\rangle = |\psi_0\rangle + \frac{1}{\varepsilon_0 - H_0} H_I |\psi_0\rangle, \tag{A.30}$$

where $|\psi_0\rangle$ is the ground state for the free fermion. The current in the presence of the applied magnetic field will then be

$$\begin{aligned}
\langle j_p \rangle &= \langle \psi_0 | j_p | \psi_0 \rangle + \langle \psi_0 | \frac{1}{\epsilon_0 - H_0} H_I | \psi_0 \rangle + \langle \psi_0 | H_I \frac{1}{\epsilon_0 - H_0} j_p | \psi_0 \rangle \\
&= -\frac{Ne^2 A_p}{mcV} - \langle \psi_0 | \sum_{k,k'} \frac{e\hbar\mu_B A_p}{2mV} \cdot (2k' - p)(2k - p) \\
&= \times \left(c^\dagger_{k-p} c_k \frac{1}{\epsilon_0 - H_0} c^\dagger_{k'} c_{k'-p} + c^\dagger_{k'} c_{k'-p} \frac{1}{\epsilon_0 - H_0} c^\dagger_{k-p} c_k \right) |\psi_0\rangle
\end{aligned} \tag{A.31}$$

A.7. Finite Temperature Quantum State Transformation

The time-dependent Schrödinger equation of a particle contains great similarity to the equations

$$H|\psi_s\rangle = i\hbar \frac{\partial}{\partial t} |\psi_s\rangle \quad \text{(dynamics)}, \tag{A.32}$$

$$H|\psi_s\rangle = -\frac{\partial}{\partial \tau} |\psi_s\rangle \quad \text{(statics)}. \tag{A.33}$$

"Time" evolved:

$$|\psi_s(\tau)\rangle = e^{-H\tau} |\psi_s(0)\rangle. \tag{A.34}$$

Heisenberg picture implies operators:

$$A_H(\tau) = e^{iH(-i\tau)} A_s e^{-iH(-i\tau)} \tag{A.35}$$

$$= e^{H\tau} A_s e^{-H\tau}. \tag{A.36}$$

Take free particle Hamiltonian:

$$H = \sum \epsilon_k c^\dagger_k c_k \Longrightarrow \text{equations of motion}, \tag{A.37}$$

$$\frac{\partial c_k}{\partial \tau} = [H, c_k] = -\epsilon_k c_k, \tag{A.38}$$

$$\frac{\partial c^\dagger_k}{\partial \tau} = [H, c^\dagger_k] = \epsilon_k c^\dagger_k. \tag{A.39}$$

Solutions are as follows:

$$c_k(\tau) = e^{-\epsilon_k \tau} c_k, \tag{A.40}$$

$$c_k(\tau)^\dagger = e^{\epsilon_k \tau} c_k^\dagger. \tag{A.41}$$

Here c_k and $c_k^\dagger$ are not Hermitian conjugates of each other.

Interaction picture (important for perturbative methods) is given by

$$|\psi_I(\tau)\rangle = e^{H_0\tau}|\psi_s(\tau)\rangle = e^{H_0\tau}e^{-H\tau}|\psi_H\rangle \tag{A.42}$$

$$= U(\tau)|\psi_H\rangle, \tag{A.43}$$

where $U(\tau)$ is time evolution operator. Also,

$$A_H(\tau) = e^{H\tau}A_s e^{-H\tau} = U^{-1}(\tau)A_I(\tau)U(\tau), \tag{A.44}$$

$$|\psi_I(\tau)\rangle = U(\tau_1)U^{-1}(\tau_2)|\psi_I(\tau_2)\rangle \tag{A.45}$$

$$= S(\tau_1,\tau_2)|\psi_I(\tau_2)\rangle, \tag{A.46}$$

where $U(\tau_1)U^{-1}(\tau_2) \equiv S(\tau_1,\tau_2)$.

Now, we need equations of motion for $U(\tau)$ and $S(\tau_1,\tau_2)$:

$$-\frac{\partial}{\partial\tau}U(\tau) = -\frac{\partial}{\partial\tau}\left[e^{H_0\tau}e^{-H\tau}\right] \tag{A.47}$$

$$= e^{H_0\tau}Ve^{-H\tau} \tag{A.48}$$

$$= e^{H_0\tau}Ve^{-H_0\tau}U(\tau) \tag{A.49}$$

$$= V_I(\tau)U(\tau). \tag{A.50}$$

Similarly,

$$-\frac{\partial}{\partial\tau}S(\tau_1,\tau_2) = V_I(\tau_1)S(\tau_1,\tau_2). \tag{A.51}$$

Formal solutions to (A.50) and (A.51) are as follows:

$$U(\tau) = \mathcal{T}e^{-\int_0^\tau V_I(\tau)d\tau}, \tag{A.52}$$

$$S(\tau_1,\tau_2) = \mathcal{T}e^{-\int_{\tau_1}^{\tau_2} V_I(\tau)d\tau}. \tag{A.53}$$

We now want to use these to calculate the partition function:

$$Z = \mathrm{Tr}[e^{-\beta H}] = \mathrm{Tr}\left[e^{-\beta H_0}U(\beta)\right] \tag{A.54}$$

$$= \text{Tr}[e^{-\beta H_0}] \cdot \frac{\text{Tr}\left[e^{-\beta H_0} U(\beta)\right]}{\text{Tr}[e^{-\beta H_0}]} \tag{A.55}$$

$$= Z_0 \times \langle U(\beta) \rangle_0. \tag{A.56}$$

This tells us that the ratio of interacting and non-interacting partition functions is given by

$$\frac{Z}{Z_0} = e^{-\beta \Delta F} = \langle \mathcal{T} e^{-\int_0^\beta V_I(\tau) d\tau} \rangle. \tag{A.57}$$

This is the main equation for finite T QFT perturbative treatment.

A.8. Matrix Product Operator Representation

Not only the quantum states in the many-body systems, but also the observable operators can be represented as the product of the matrices (or the lower rank tensors) solely related to each of one-sites, which can be called the *matrix product operator* (MPO). This disassembly of the operators helps the dimension of the N-particle systems to be reduced from the exponential order to the linear order, which is capable of approximately calculating the physical properties of many-body systems such as the ground state and the two-point correlation.

Similar to the matrix product state, the many-body operator $\hat{\mathcal{O}}$ for the open boundary condition takes the form

$$\mathcal{O}^{(s_1,s_1'),(s_2,s_2'),\ldots,(s_n,s_n')} = \sum_{\beta_1,\ldots,\beta_N}^{D_O} (O^{[1]})_{\beta_1}^{s_1,s_1'} (O^{[2]})_{\beta_1,\beta_2}^{s_2,s_2'} \cdots (O^{[n]})_{\beta_{n-1}}^{s_n,s_n'}, \tag{A.58}$$

where s_i is the physical index which mainly means a basis of the z-directional spin at ith site and β_i is the auxiliary index between two auxiliary parties with i index [159]. The s_i ranges from 1 to d with d the dimension of local Hilbert space and the β_i is in the range of 1 to D_O called the auxiliary dimension. The rank of $O^{[i]}$ is four except two tensors $O^{[1]}$ and $O^{[N]}$ for the open boundary condition. For the periodic boundary systems, however, these two tensors become four-rank tensors by adding auxiliary index β_N to each tensor (Fig A.1).

The methods to disassemble a high-rank tensor into a product of the matrices (or the lower-rank tensors) depending only on the one-site are suggested [164–166]. Before touching the 1D Hamiltonians for the various interactions, let us illustrate the simplest case: the Hamiltonian for the 1D

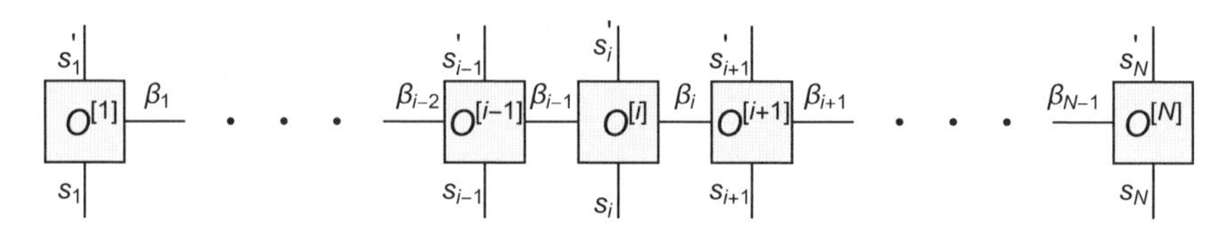

Fig. A.1. The matrix product operator diagram for the open boundary condition.

spin-1/2 Ising model with the only nearest neighbor interaction

$$\hat{\mathcal{H}} = \sum_{i=1}^{N-1} \hat{\sigma}_i^z \otimes \hat{\sigma}_{i+1}^z \tag{A.59}$$

$$= O^{[1]}O^{[2]}\cdots O^{[N]}, \tag{A.60}$$

where $\hat{\sigma}_i^z$ is the z-directional Pauli operator on ith site and

$$O^{[i]} \equiv O = \begin{pmatrix} \mathbb{1} & \hat{\sigma}_i^z & 0 \\ 0 & 0 & \hat{\sigma}_i^z \\ 0 & 0 & \mathbb{1} \end{pmatrix} \tag{A.61}$$

with $2 \leq i \leq N-1$ and the boundary matrices are $O^{[1]} = (\mathbb{1}, \hat{\sigma}_1^z, 0)$ and $O^{[N]} = (0, \hat{\sigma}_N^z, \mathbb{1})^T$ [166]. Henceforth, the decomposed rank-4 tensor $(O^{[i]})_{\beta_{i-1},\beta_i}^{s_i,s_i'}$ is regarded as the matrix $O^{[i]}_{(\beta_{i-1}s_i)(\beta_i s_i')}$ whose inner part (e.g. $\hat{\sigma}_i^z$ and $\mathbb{1}$ in Eq. (A.61)) and outer part correspond to the physical indices (s_i, s_i') and the auxiliary indices (β_{i-1}, β_i), respectively.

It is the general description that explains how to obtain the decomposed tensors (or matrices) for the Hamiltonians not just with the nearest neighbor interaction, but also with all kinds of two-body interactions. According to Fröwis' paper [166], the decomposed matrix $O^{[i]}$ in Eq. (A.61) has the auxiliary indices β_{i-1} and β_i, called *left* and *right inputs*, which are connected to the right input of the nearest matrix $O^{[i-1]}$ and the left input of $O^{[i+1]}$, respectively. Some combinations of left and right inputs lead to the operators, others lead to a zero operator. For instance, in Eq. (A.61), the input $(\beta_{i-1}, \beta_i) = (1, 2)$ corresponds to $\hat{\sigma}_i^z$. Each possible combination of left and right inputs indicating a non-zero operator is called a *rule*.

Let us exemplify finding the rules for the Hamiltonian of Eq. (A.59) which consists of the sum of the strings $\mathbb{1}^{\otimes(i-1)} \otimes \hat{\sigma}_i^z \otimes \hat{\sigma}_{i+1}^z \otimes \mathbb{1}^{\otimes(N-(i+1))}$.

Shortly, the expression of this sting is $\mathbb{1} \otimes \hat{\sigma}^z \otimes \hat{\sigma}^z \otimes \mathbb{1}$. If the two tensors adjoin each other, the right input of the left tensor and the left input of the right tensor in this string are the same. Then, one can write down a set of rules as Table A.1, where the first and the last identity operators $\mathbb{1}$ in this string must be the first and last diagonal elements of the each decomposed matrices, respectively. Finally, the combination of all rules results in the decomposed matrix equation (A.61).

In the same manner, one can obtain the MPO representations for the case of the more complicated two-body interactions by finding the family of rules. Let us consider the Ising model with nearest neighboring and next-nearest neighboring interactions:

$$\hat{\mathcal{H}} = \sum_{i=1}^{N-1} J_1 \hat{\sigma}_i^x \otimes \hat{\sigma}_{i+1}^x + \sum_{i=1}^{N-2} J_2 \hat{\sigma}_i^x \otimes \hat{\sigma}_{i+2}^x + \sum_{i=1}^{N} h \hat{\sigma}_i^z. \qquad \text{(A.62)}$$

The family of rules for the Hamiltonian equation (A.62) shown in Table A.2 can be obtained [166]. Compared to Table A.1, the new rules 2, 3

Table A.1. Family of rules for the Hamiltonian equation (A.59).

Rule number	Input(left, right)		Output
1	$(1,1)$	$\rightarrow$	$\mathbb{1}$
2	$(1,2)$	$\rightarrow$	$\hat{\sigma}^z$
3	$(2,3)$	$\rightarrow$	$\hat{\sigma}^z$
4	$(3,3)$	$\rightarrow$	$\mathbb{1}$

Table A.2. Set of rules for the Hamiltonian equation (A.62).

Rule number	Input(left, right)		Output
1	$(1,1)$	$\rightarrow$	$\mathbb{1}$
2	$(1,2)$	$\rightarrow$	$\hat{\sigma}^x$
3	$(2,3)$	$\rightarrow$	$\mathbb{1}$
4	$(3,4)$	$\rightarrow$	$\hat{\sigma}^x$
5	$(2,4)$	$\rightarrow$	$\hat{\sigma}^x$
6	$(4,4)$	$\rightarrow$	$\mathbb{1}$
7	$(1,4)$	$\rightarrow$	$\hat{\sigma}^z$

corresponding to the next-nearest neighboring interactions are added in Table A.2. These rules in Table A.2 facilitate the construction of the decomposed matrices:

$$O^{[i]} \equiv O = \begin{pmatrix} \mathbb{1} & \hat{\sigma}^x & 0 & h\hat{\sigma}^z \\ 0 & 0 & \mathbb{1} & J_1\hat{\sigma}^x \\ 0 & 0 & 0 & J_2\hat{\sigma}^x \\ 0 & 0 & 0 & \mathbb{1} \end{pmatrix} \quad \text{with } 2 \le i \le N-1, \tag{A.63}$$

$$O^{[1]} = \begin{pmatrix} \mathbb{1} & \hat{\sigma}^x & 0 & \hat{\sigma}^z \end{pmatrix} \quad \text{and} \quad O^{[N]} = \begin{pmatrix} h\hat{\sigma}^z & J_1\hat{\sigma}^x & J_2\hat{\sigma}^x & \mathbb{1} \end{pmatrix}^T. \tag{A.64}$$

A.9. The Variational Principle in the MPS Representation

If there is a Hamiltonian $\hat{\mathcal{H}}$ for the open boundary condition and we want to find its ground energy and the corresponding ground state, one way is to use the variational method which means optimizing the expectation value $\langle\psi|\hat{\mathcal{H}}|\psi\rangle$ with constraint $\langle\psi|\psi\rangle = 1$, that is,

$$\frac{\partial}{\partial\langle\psi|}(\langle\psi|\hat{\mathcal{H}}|\psi\rangle - \mathcal{E}\langle\psi|\psi\rangle) = 0, \tag{A.65}$$

where $\mathcal{E}$ is the Lagrangian multiplier. Instead of the quantum state $\langle\psi|$, the ith matrix in any MPS form is regarded as a variable $M^{[i]}$. Then the variational method becomes

$$\frac{\partial}{\partial M^{\dagger[i]}}(M^{\dagger[i]}\hat{H}^{[i]}_{\text{eff}}M^{[i]} - \mathcal{E}M^{\dagger[i]}\mathbb{1}^{[i]}_{\text{eff}}M^{[i]}) = 0, \tag{A.66}$$

$$\hat{H}^{[i]}_{\text{eff}}M^{[i]} - \mathcal{E}\mathbb{1}^{[i]}_{\text{eff}}M^{[i]} = 0, \tag{A.67}$$

where $\hat{H}^{[i]}_{\text{eff}}$ (or $\mathbb{1}^{[i]}_{\text{eff}}$) in Fig. A.2 consists of the summation of the class of matrices for the ground state $|\psi\rangle$ and the operator $\hat{\mathcal{H}}$ (or $\mathbb{1}$) except

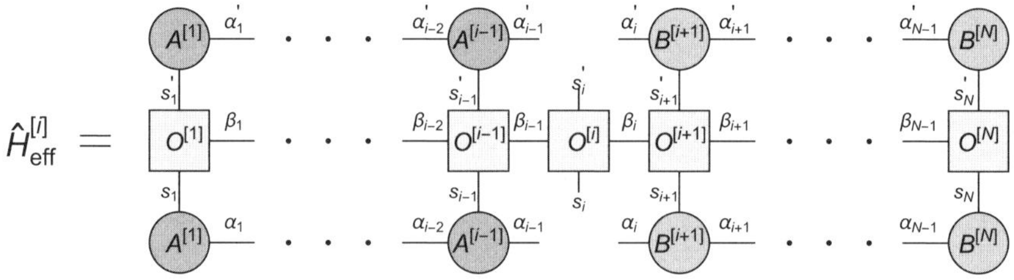

$$\hat{H}^{[i]}_{\text{eff}} =$$

Fig. A.2. The tensor network diagram of $\hat{H}^{[i]}_{\text{eff}}$: the six-rank tensor is expressed as matrix elements $H_{(\alpha'_{i-1}s'_i\alpha'_i)(\alpha_{i-1}s_i\alpha_i)}$.

the ith-site matrix of $|\psi\rangle$ and its conjugate transpose. Equation (A.67) is the generalized eigenvalue problem [181, 189]. In the case of the open boundary condition, $\mathbb{1}_{\text{eff}}^{[i]}$ becomes the identity operator, where Eq. (A.67) is the eigenvalue problem. However, the problem is that the dimension of the matrices $\hat{H}_{\text{eff}}^{[i]}$ and $\mathbb{1}_{\text{eff}}^{[i]}$ is too large to find all the eigenvalues and their corresponding eigenvectors. Since the only thing we want is the ground state, it is sufficient to use the Lanczos algorithm to obtain the lowest eigenvalue with its eigenvector for the large sparse matrix [168]. Iteratively solving Eq. (A.67) for all $i \leq N$, one can obtain all ith-site matrices and the many-body ground state is completed. The specific algorithm [159] is shown in Table A.3 with the diagrams of its tensor network.

Table A.3. The MPS algorithm of the variational principle.

(i)	Prepare the initial random right-normalized matrices $B^{[i]}$ in Eq. (4.88) or the initial ground state from the infinite-size DMRG method
(ii)	Construct $\hat{H}_{\text{eff}}^{[1]}$ and $\mathbb{1}_{\text{eff}}^{[1]}$. Then, by using the Lanczos algorithm, solve the generalized eigenvalue problem (A.67) to find the new matrix $M^{[1]}$
(iii)	Apply the SVD to the matrix $M^{[1]}$, that is, $M^{[1]} = U\Lambda V^{\dagger}$. Then, replace the matrix $A^{[1]}$ with $\tilde{A}^{[1]} := U$
(iv)	Construct the $\hat{H}_{\text{eff}}^{[2]}$ and $\mathbb{1}_{\text{eff}}^{[2]}$ with the updated matrix $\tilde{A}^{[1]}$. Again, solve the generalized eigenvalue problem and find the updated matrix $\tilde{A}^{[2]}$. Repeat this step until $\tilde{A}^{[N]}$ is obtained
(v)	At this time, start to solve the generalized eigenvalue problem for $i = N$ involving all matrices $\tilde{A}$. After calculating the SVD of $M^{[N]}$, i.e. $M^{[N]} = U\Lambda V^{\dagger}$, update the matrix $\tilde{B}^{[N]} := V^{\dagger}$
(vi)	Repeat the step (v) up to $i = 1$ and obtain all updated matrices $\tilde{B}$
(vii)	Iterate the steps (i)–(vi) until the ground state converges

Bibliography

[1] P. P. Ewald, *Fifty Years of X-ray Diffractions* (N. V. A. Oosthoeâs Uitgeversmaatschappij, Utrecht, 1962).

[2] N. W. Ashcroft and N. D. Mermin, in Solid State Physics (Holt, Rinehart and Winston, New York, 1976), p. 8.

[3] G. Duyckaerts, *Physica* **6**, 817 (1939).

[4] W. S. Corak *et al.*, *Phys. Rev.* **98**, 1699 (1955).

[5] A. J. Walton, *Three Phases of Matter* (Clarendon Press, Oxford, 1983).

[6] M. P. Marder, *Condensed Matter Physics* (Wiley-Interscience, New York, 1960).

[7] M. Ali Omar, *Elementary Solid State Physics* (Addison-Wesley, Reading, MA, 1975).

[8] A. H. Wilson, Opportunities lost and opportunities seized, in *The Beginnings of Solid State Physics* (Royal Society, London, 1980), pp. 39–48.

[9] R. G. Parr and W. Yang, *Density-Functional Theory of Atoms and Molecules* (Clarendon Press, Oxford, 1989).

[10] P. A. M. Dirac, *Proc. Roy. Soc. London A* **123**, 714 (1929).

[11] P. A. M. Dirac, in *The Principles of Quantum Mechanics*, 4th edn. (Oxford University Press, Oxford, 1958), p. 248.

[12] G. D. Mahan, *Many-Particle Physics* (Plenum Press, New York, 1990).

[13] S. M. Barnett and P. M. Radmore, *Methods in Theoretical Quantum Optics*, (Clarendon Press, Oxford, 1997).

[14] P. Phillips, *Advanced Solid State Physics* (Westview Press, Boulder, 2003).

[15] L. D. Landau and E. M. Lifshitz, *Statistical Physics* (Elsevier, 1951).

[16] C. Cohen-Tannoudji, B. Diu and F. Laloë, *Quantum Mechanics*, 2nd edn. (J. Wiley, New York, 1977).

[17] E. O'Reilly, *Quantum Theory of Solids* (Taylor & Francis, 2002).

[18] J. M. Yeomans, *Statistical Mechanics of Phase Transitions* (Clarendon Press, Oxford, 1992).

[19] J. D. Patterson and B. C. Bailey, *Solid-State Physics* (Springer, Berlin, 2007).
[20] L. D. Landau and E. M. Lifshitz, *Statistical Physics Part* 1, 3rd edn., Course of Theoretical Physics, Vol. 5 (Pergamon Press, Oxford, 1994).
[21] O. Penrose, *Phil. Mag.* **42**, 1373 (1951).
[22] J. Bardeen, L. N. Cooper and J. R. Schrieffer, *Phys. Rev.* **108**, 1175 (1957).
[23] C. N. Yang, *Rev. Mod. Phys.* **34**, 694 (1962).
[24] R. Peierls, *Proc. Cambridge Philos Soc.* **32**, 477 (1936); R. B. Griffiths, *Phys. Rev. A* **136**, 437 (1964).
[25] N. H. Mermin and H. Wagner, *Phys. Rev. Lett.* **17**, 1133 (1966).
[26] P. C. Hohenberg, *Phys. Rev.* **158**, 383 (1967).
[27] V. L. Berezinskii, *Zh. Eksp. Teor. Fiz.* **59**, 907 (1970); [*Sov. Phys. JEPT* **32**, 493 (1971)]; J. M. Kosterlitz and D. J. Thouless, *J. Phys.* **6**, 1181 (1973).
[28] S. Sachdev, *Quantum Phase Transitions* (Cambridge University Press, Cambridge, 2000).
[29] A. Osterloh, L. Amico, G. Falci and R. Fazio, *Nature (London)* **416**, 608 (2002).
[30] L. Amico, R. Fazio, A. Osterloh and V. Vedral, *Rev. Mod. Phys.* **80**, 517 (2007).
[31] H. T. Nieh, G. Su and B. H. Zhao, *Phys. Rev. B* **51**, 3760 (1994).
[32] G. Su and M. Suzuki, *Phys. Rev. Lett.* **86**, 2708 (2001).
[33] T. D. Kühner, S. R. White and H. Monien, *Phys. Rev. B* **61**, 12474 (2000).
[34] V. Vedral, *Nero. J. Phys.* **6**, 102 (2004).
[35] B. S. Shastry and B. Sutherland, *Phys. Rev. B* **47**, 7995 (1993).
[36] B. S. Shastry and B. Sutherland, *Phys. Rev. Lett.* **65**, 243 (1990).
[37] A. M. Rey *et al.*, *J. Phys. B: Atom. Mol. Opt. Phys.* **36**, 825 (2003).
[38] M. T. Cuhna, J. Dunningham and V. Vedral, *Proc. Roy. Soc. A* **463**, 2277 (2007).
[39] V. Vedral, quant-ph/0410021.
[40] M. C. Arnesen, S. Bose and V. Vedral, *Phys. Rev. Lett.* **87**, 017901 (2001).
[41] Č. Brukner, V. Vedral and A. Zeilinger, *Phys. Rev. A* **73**, 012110 (2006).
[42] Č. Brukner and V. Vedral, quant-ph/040604 (2004).
[43] J. Hide *et al.*, quant-ph/0701013.
[44] G. Toth, *Phys. Rev. A* **71**, 010301(R) (2005).
[45] S.-J. Gu, G.-S. Tian and H.-Q. Lin, *Phys. Rev. A* **71**, 052322 (2005).
[46] M. Bortz and Göhmann, *Eur. Phys. J. B* **46**, 399 (2005).
[47] A. Polkovnikov, E. Altman and E. Demler, *Proc. Natl. Acad. Sci. USA* **103**, 6125 (2006).
[48] Z. Hadzibabic *et al.*, *Nature* **441**, 1118 (2006).
[49] S. Ghosh, T. F. Rosenbaum, G. Aeppli and S. N. Coppersmith, *Nature* **425**, 48 (2003).
[50] E. Fradkin, *Phys. Rev. Lett.* **63**, 322 (1989).
[51] O. Derzhko *et al.*, *Physica A* **320**, 407 (2003).
[52] M. A. Nielsen and I. L. Chuang, *Quantum Computation and Quantum Information* (Cambridge University Press, Cambridge, 2000).

[53] V. Vedral, *Nature* **425**, 28 (2008).
[54] O. Penrose and L. Onsager, *Phys. Rev.* **104**, 576 (1956).
[55] V. Vedral, *Rev. Mod. Phys.* **74**, 197 (2002).
[56] V. Vedral, M. B. Plenio, M. A. Rippin and P. L. Knight, *Phys. Rev. Lett.* **78**, (1997).
[57] V. Vedral and M. B. Plenio, *Phys. Rev. A* **57**, 1619 (1998).
[58] T.-C. Wei, M. Ericcson, P. M. Goldbart and W. J. Munro, quant-ph 0405002.
[59] M. B. Plenio and V. Vedral, *J. Phys. A* **34**, 6997 (2001).
[60] C. N. Yang, *Phys. Rev. Lett.* **63**, 2144 (1989).
[61] N. M. Plakida, *High Temperature Superconductivity: Experiment and Theory* (Springer, New York, 1995).
[62] P. Zanardi and X. G. Wang, *J. Phys. A* **35**, 7947 (2002).
[63] H. Fan, Private communication, December (2003).
[64] V. Vedral, *Cen. Eur. J. Phys.* **2**, 289 (2003).
[65] Y. Shi, *Phys. Rev. A* **67**, 024301 (2003).
[66] K. Eckert, J. Schliemann, D. Bruss and M. Lewenstein, *Ann. Phys.* **299**, 88 (2002).
[67] F. H. L. Essler, V. E. Korepin and K. Schoutens, *Phys. Rev. Lett.* **68**, 2960 (1992).
[68] S. Schneider and G. J. Milburn, *Phys. Rev. A* **65**, 042107 (2002); X. G. Wang and C. Molmer, *Eur. Phys. J. D* **18**, 385 (2002).
[69] A. Peres, *Phys. Lett. A* **202**, 16 (1995).
[70] L. Henderson and V. Vedral, *J. Phys. A* **34** (2001); V. Vedral, *Phys. Rev. Lett.* **90**, 050401 (2003).
[71] K. G. H. Vollbrecht and R. F. Werner, *Phys. Rev. A* **64**, 062307 (2001).
[72] P. Zanardi and X. G. Wang, *J. Phys. A* **35**, 7947 (2002); H. Fan and S. Lloyd, quant-ph/0405130.
[73] K. Eckert, J. Schliemann, D. Bruss and M. Lewenstein, *Ann. Phys.* **299**, 88 (2002).
[74] Y. Aharonov and D. Bohm, *Phys. Rev.* **115**, 485 (1959).
[75] S. Weinberg, *The Quantum Theory of Fields: Vol. 2, Modern Applications* (Cambridge University Press, Cambridge, 2000).
[76] P. W. Higgs, *Phys. Rev. Lett.* **13**, 508 (1964).
[77] P. Calabrese and J. Cardy, *J. Stat. Mech.* P06002 (2004).
[78] H. Casini and M. Huerta, hep-th/0405111.
[79] S. Oh and J. Kim, quant-ph/0406013.
[80] P. Zanardi and M. Rasetti, *Phys. Lett. A* **264**, 94 (1999).
[81] C. Nayak, S. H. Simon, A. Stern, M. Freedman and S. Das Sarma, *Rev. Mod. Phys.* **80**, 1083 (2008).
[82] Y. Nakamura, Yu. Pashkin and J. S. Tsai, *Nature* **398**, 786 (1999).
[83] G. Falci, R. Fazio, G. M. Palma, J. Siewert and V. Vedral, *Nature* **407**, 21 (2000).
[84] P. J. Leek, J. M. Fink, A. Blais, R. Bianchetti, M. Gppl, J. M. Gambetta, D. I. Schuster, L. Frunzio, R. J. Schoelkopf and A. Wallraff, *Science* **318**, 21 (2007).

[85] M. Tinkham, *Introduction to Superconductivity*, 2nd edn. (McGraw-Hill, New York, 1996).

[86] R. Fazio, M. G. Palma and J. Siewert *Phys. Rev. Lett.* **83**, 5385 (1999).

[87] D. Castelvecchi, *Nature* **541**, 9 (2017).

[88] https://www-03.ibm.com/press/us/en/pressrelease/49661.wss.

[89] https://newsroom.intel.com/news-releases/intel-invests-us50-million-to-advance-quantum-computing/.

[90] L. Onsager, *Phys. Rev.* **65**, 117 (1944).

[91] E. H. Lieb, in *Proc.* 1993 *Con. Honor of G.F. Dell'Antonio, Advances in Dynamical Systems and Quantum Physics* (World Scientific, Singapore, 1995), pp. 173–193.

[92] J. Hubbard, *Proc. Roy. Soc. London A* **276**, 238 (1963).

[93] I. Bloch and M. Greiner, *Adv. At. Mol. Opt. Phys.* **52**, 1 (2005).

[94] M. Lewenstein *et al.*, *Adv. Phys.* **56**, 243 (2007).

[95] D. Jaksch and P. Zoller, *Ann. Phys.* **315**, 52 (2005).

[96] M. P. A. Fisher, P. B. Weichman, G. Grinstein and D. S. Fisher, *Phys. Rev. B* **40**, 546 (1989).

[97] J. Kondo, *Prog. Theoret. Phys.* (*Kyoto*) **32**, 37 (1964).

[98] L. Kouwenhoven and L. Glazman, *Phys. World* **14**(1), 33–38 (2001).

[99] P. W. Anderson, G. Yuval and D. R. Hamann, *Phys. Rev. B* **1**, 4464 (1970).

[100] P. W. Anderson, *Phys. Rev.* **124**, 41 (1961).

[101] G. Grosso and G. P. Parravicini, *Solid State Physics* (Academic Press, London, 2000).

[102] J. R. Schrieffer and P. A. Wolff, *Phys. Rev.* **149**, 491 (1966).

[103] A. Einstein, B. Podolsky and N. Rosen, *Phys. Rev.* **47**, 777 (1935).

[104] D. Loss and D. P. DiVincenzo, *Phys. Rev. A* **57**, 120 (1998).

[105] A. Imamoglu, D. D. Awschalom, G. Burkard, D. P. DiVincenzo, D. Loss, M. Sherwin and A. Small, *Phys. Rev. Lett.* **83**, 4204 (1999).

[106] D. P. Divincenzo *et al.*, *Nature* **408**, 339 (2000).

[107] X. Wang, *Phys. Rev. A* **64**, 012313 (2001).

[108] M. C. Arnesen, S. Bose and V. Vedral, *Phys. Rev. Lett.* **87**, 017901 (2001).

[109] C. Brukner and V. Vedral, quant-ph/0406040.

[110] M. R. Dowling, A. C. Doherty and S. D. Bartlett, *Phys. Rev. A* **70**, 062113 (2004).

[111] L. A. Wu, S. Bandyopadhyay, M. S. Sarandy and D. A. Lidar, *Phys. Rev. A* **72**, 032309 (2005).

[112] M. Oshikawa and I. Affleck, *Phys. Rev. Lett.* **79**, 2883 (1997).

[113] H. Nojiri, Y. Ajiro, T. Asano and J. P. Boucher, *New J. Phys.* **8**, 218 (2006).

[114] M. Wiesniak, V. Vedral and C. Brukner, *New J. Phys.* **7**, 258 (2005).

[115] M. Suzuki, *J. Phys. Soc. Japan* **21**, 2140 (1966).

[116] S. Katsura, *Phys. Rev.* **127**, 1508 (1962).

[117] S. Katsura, *Phys. Rev.* **129**, 2835 (1963).

[118] M. Horodecki, P. Horodecki and R. Horodecki, *Phys. Lett. A* **223**, v1 (1996).

[119] W. Heitler and F. London, *Z. Phys.* **44**, 455 (1927).

[120] W. Heisenberg, *Z. Phys.* **38**, 411 (1926).

[121] P. A. M. Dirac, *Proc. Roy. Soc. London A* **112**, 661 (1926).
[122] R. J. Baxter, *Exactly Solved Models in Statistical Mechanics* (Academic Press Inc., London, 1982).
[123] K. Huang, *Statistical Mechanics* (J. Wiley, New York, 1987).
[124] A. J. Leggett, *Rev. Mod. Phys.* **71**, S318 (1999).
[125] C. A. Reynolds, B. Serin, W. H. Wright and L. B. Nesbitt, *Phys. Rev.* **78**, 487 (1950).
[126] W. Meissner and R. Ochsenfeld, *Naturwissenschaften* **21**, 787 (1933).
[127] H. Frohlich, *Proc. Phys. Soc. London A* **63**, 778 (1950).
[128] L. N. Cooper, *Phys. Rev.* **104**, 1189 (1956).
[129] J. Bardeen, L. N. Cooper and J. R. Schrieffer, *Phys. Rev.* **106**, 162 (1957).
[130] P. L. Taylor and O. Heinonen, *A Quantum Approach to Condensed Matter Physics* (Cambridge University Press, Cambridge, 2002).
[131] L. Boltzmann, *Lectures on Gas Theory* (University of California Press, Berkley, CA, 1964).
[132] G. M. Kremer, *An Introduction to the Boltzmann Equation and Transport Processes in Gases* (Springer, Berlin, 2009).
[133] D. Snoke, *Ann. Phys.* **523**, 87 (2010).
[134] D. Snoke, in *Solid State Physics: Essential Concepts*, Chapter 4.8 (Pearson/Addison-Wesley, San Francisco, 2009).
[135] I. Bengtsson and K. Zyczkowski, *Geometry of Quantum States: An Introduction to Quantum Entanglement* (Cambridge University Press, New York, 2006).
[136] D. Bruss, *J. Math. Phys.* **43**, 4237 (2002).
[137] J. Eisert, Entanglement in quantum information theory, Ph.D. thesis, University of Potsdam (2001).
[138] R. Horodecki, P. Horodecki, M. Horodecki and K. Horodecki, *Rev. Mod, Phys.* **81**, 865 (2009).
[139] M. B. Plenio and V. Vedral, *Contemp. Phys.* **39**, 431 (1998).
[140] M. B. Plenio, *Quantum Inf. Comput.* **7**, 1 (2007).
[141] W. K. Wootters, *Quantum Inf. Comput.* **1**, 27 (2001).
[142] S. Sachdev, *Quantum Phase Transitions* (Cambridge University Press, Cambridge, 1999).
[143] M. Takahashi, *Thermodynamics of One Dimensional Solvable Models* (Cambridge University Press, Cambridge, 1999).
[144] E. Fradkin and L. Susskind, *Phys. Rev. D* **17**, 2637 (1978).
[145] L. Amico *et al.*, *Rev. Mod. Phys.* **80**, 517 (2008).
[146] H. T. Quan *et al.*, *Phys. Rev. Lett.* **96**, 140604 (2006); P. Zanardi and N. Paunkovic, *Phys. Rev. E* **74**, 031123 (2006).
[147] V. Bužek, M. Orszag and M. Roko, *Phys. Rev. Lett.* **94**, 163601 (2005).
[148] A. De Pasquale *et al.*, *Eur. Phys. J. Spec. Top.* **160**, 127 (2008); arxiv: 0801.1394.
[149] E. Lieb, T. Schultz and D. Mattis, *Ann. Phys.* **16**, 407 (1961).
[150] W. K. Wootters, *Phys. Rev. Lett.* **80**, 2245 (1998).

[151] N. Paunkovic *et al.*, *Phys. Rev. A* **77**, 052302 (2008); H. Kwok, C. Ho and S. Gu, quant-ph/0805.3885.
[152] F. Franchini *et al.*, *J. Phys. A* **40**, 8467 (2007).
[153] H.-J. Mikeska and W. Pesch, *Z. Phys. B* **26**, 351 (1977).
[154] T. Tonegawa, *Solid State Comm.* **40**, 983 (1981).
[155] B.-Q. Jin and V. E. Korepin, *Phys. Rev. A* **69**, 062314 (2004).
[156] L. Amico *et al.*, *Phys. Rev. A* **74**, 022322 (2006).
[157] H.-J. Mikeska, S. Miyashita and G. H. Ristow, *J. Phys. C* **3**, 2985 (1990).
[158] H. J. Briegel and R. Raussendorf, *Phys. Rev. Lett.* **86**, 910 (2001).
[159] U. Schollwöck, *Ann. Phys.* **326**, 96 (2011).
[160] G. Vidal, *Phys. Rev. Lett.* **91**, 147902 (2003).
[161] F. Verstraete, D. Porras and J. I. Cirac, *Phys. Rev. Lett.* **93**, 227205 (2004).
[162] F. Verstraete and J. I. Cirac, arXiv:cond-mat/0407066.
[163] S. R. White, *Phys. Rev. Lett.* **69**, 2863 (1992).
[164] G. M. Crosswhite and D. Bacon, *Phys. Rev. A* **78**, 012356 (2008).
[165] G. M. Crosswhite, A. C. Doherty and G. Vidal, *Phys. Rev. B* **78**, 035116 (2008).
[166] F. Fröwis, V. Nebendahl and W. Dür, *Phys. Rev. A* **81**, 062337 (2010).
[167] R. Orús, *Ann. Phys.* **33**, 349 (2013).
[168] B. N. Parlett, *The Symmetric Eigenvalue Problem* (SIAM, Philadelphia, 1998).

Index

C

D

E

F

W

X

Z